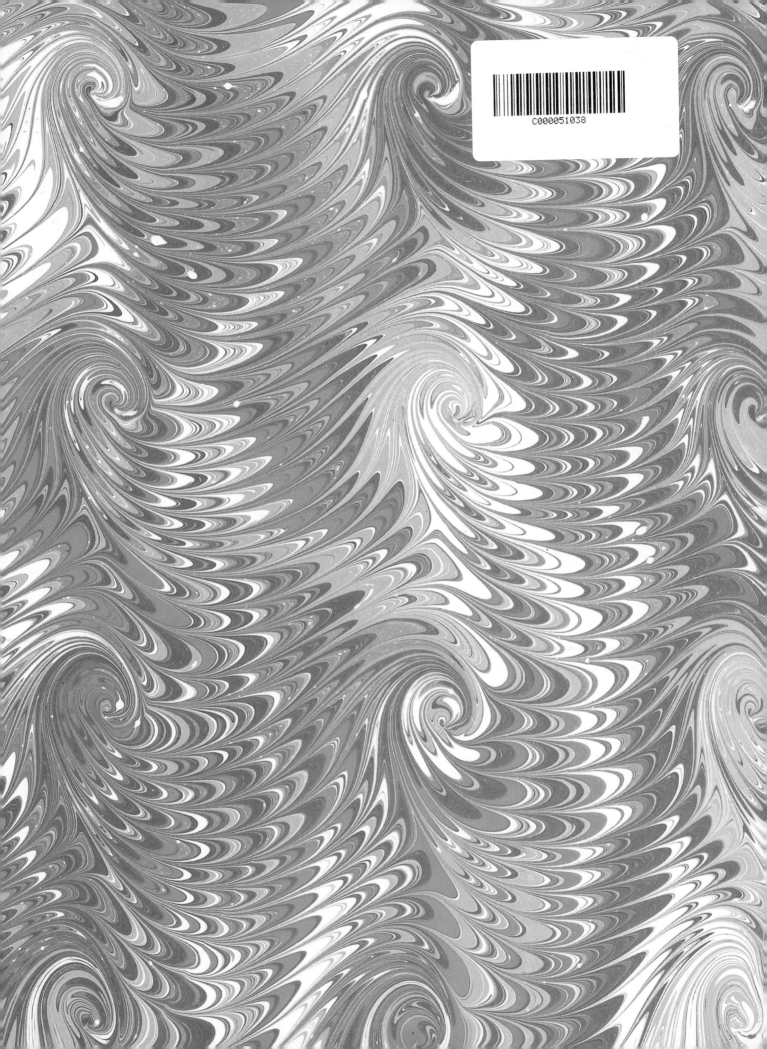

FEAST FOR THE EYES

OMEGA DESIGNS

REVISED 2ND EDITION

ANTON KREUZER

4880 Lower Valley Road, Atglen, Pennsylvania 19310

Illustration opposite page:
The famous astronaut watch Speedmaster Professional by Omega, which can be purchased at any authorized dealer for a reasonable price. This classic watch is a unique timepiece in a stainless steel case with a water-resistant-type screwed back, integrated flexible stainless steel bracelet, the bezel calibrated for tachometer for measuring speeds from 60 to 500 kilometers per hour and the black matte dial with easy readable calibrations. Its heart is a mechanical movement with manual winding. Since 1965 this watch has been NASA's official wristwatch. Since 1975, it has also been worn by Russian cosmonauts. In addition, the model is water-resistant to a depth of 30 meters.

Schiffer Books are available at special discounts for bulk purchases for sales promotions or premiums. Special editions, including personalized covers, corporate imprints, and excerpts can be created in large quantities for special needs. For more information contact the publisher.
Published by Schiffer Publishing Ltd.
4880 Lower Valley Road
Atglen, PA 19310
Phone: (610) 593-1777; Fax: (610) 593-2002
E-mail: Info@schifferbooks.com

For the largest selection of fine reference books on this and related subjects, please visit our web site at
www.schifferbooks.com
We are always looking for people to write books on new and related subjects. If you have an idea for a book please contact us at the above address.
This book may be purchased from the publisher.
Include $3.95 for shipping.
Please try your bookstore first.
You may write for a free catalog.
In Europe, Schiffer books are distributed by
Bushwood Books
6 Marksbury Ave.
Kew Gardens
Surrey TW9 4JF England
Phone: 44 (0) 20 8392-8585; Fax: 44 (0) 20 8392-9876
E-mail: info@bushwoodbooks.co.uk
Website: www.bushwoodbooks.co.uk
Free postage in the U.K., Europe; air mail at cost.

English edition copyright © 1996 and 2008 Schiffer Publishing, Ltd.
Translated from the German by Gertraud Hechl
Originally published as *Omega Modelle, Augenweide Armbanduhr*
by Carintha Verlag, Klagenfurt
Library of Congress Control Number: 2008921560

ISBN: 978-0-7643-2995-1
Printed in China

Contents

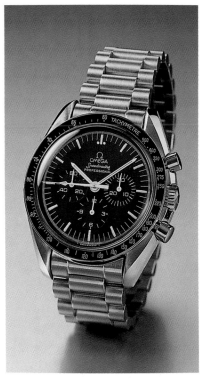

Excerpt from an Omega catalogue from 1931. The illustration shows a view of the Factory and the most important medals awarded to the company since 1889.

Introduction

The book series WRISTWATCHES-FEAST FOR THE EYES attempts to convey basic information through text and illustrations. Its purpose is not to give comprehensive information, but rather a quick overview. An introduction to the fascinating collector's field of wristwatches is given in the book SWISS MAKERS, a well illustrated lexicon of the 250 best known makers, manufacturers, and trade names and their most important technical achievements and watch creations. For the collector specializing in a certain maker, books on PATEK PHILIPPE, ROLEX, and VACHERON CONSTANTIN have been published so far.

The following book is not about a prestigious luxury watchmaker, but about a world-famous company with a long history, the OMEGA company. In 1994 the 100th anniversary of Omega's registration as a watchmaker was celebrated.

The owner of the Omega trade name is the Bieler company Louis Brandt & Freres, whose products were sold under the Omega name since 1894 and have become well known. The company was the initiator of the name consciousness of watch purchasers, which was passed on from generation to generation. If the father owned an Omega, so did the son.

The only thing left of this once enormous company is its trade name. The famous company, which since the 1930s also made its name through time keeping at sports events and as official time keeper of numerous summer and winter Olympics, is now part of the Holding SMH (Swiss corporation for micro electronics and watchmaking). It stopped producing its own cases and movements and buys movements—just like most other watch companies—from ETA, the world's largest supplier of mechanical and electronic watch calibers. Omega's largest source of pride continues to be the astronaut chronograph watch Speedmaster Professional, the chronometer wristwatch Constellation, and the diver's watch Seamaster, and the luxury line Louis Brandt, made since the 1980s in memory of its founder Louis Brandt, and a new series of mechanical luxury watches including a chronometer calendar watch, a chronograph watch with phases of the moon, and a perpetual calendar wristwatch, made since 1990.

5

The Wristwatches by Omega

The brothers Louis-Paul and Cesar Brandt, who continued their father's work, were not only gifted salesmen but also talented entrepreneurs who made their watch manufacturer the largest in Switzerland within a few years. Spurred by the vision to produce pocket watches of highest quality in an industrial setting, they manufactured a pocket watch from 1894 on by using the most modern manufacturing methods and the latest machinery, which was considered nonplus at its time. Hence, the company was named Omega, the last letter of the Greek alphabet, a synonym for the highest achievement without equal. The graphic sign was chosen at the same time as the new trademark. Over the years the symbol has achieved unbelievable publicity and the company has become the flag ship of the Swiss watchmaking industry.

The Brandts were apparently fascinating personalities, full of visions and ideas. One day, they themselves or an acquaintance had the idea to add a second pendant to a minute repeating pocket watch at six o'clock and to use those to attach the watch to the wrist. It is unknown why the watch was attached to the wrist. It is conceivable that it was originally designed for a competition or an exhibition. It is also unknown what the monogram "LLHMS" stands for on the outer case, as is the meaning of the inscription on the cuvette. The inscription reads "Hors Concours. Membre de Jury. Paris 1889—L. Brandt & Freres—Bienne (Suisse)—1, Rue d'Hauteville, Paris" Is it a commemorative piece from the world exhibition in Paris? Cesar Brandt, who since 1888 was heading a sales office in Paris, had been named by the Swiss administration to be a member of the international jury, watchmaking division. This honorary nomination, however, excluded the firm from participating in the competition.

Left: A custom made Gold Minute Repeating Watch. By attaching a second pendant a small pocket watch was converted into a watch, which could be worn on the wrist.

7

Early wristwatch by Omega. The gentleman's watch is impressesive with its simple silver hinged case and white enamel dial with large, easy to read numerals. The problem of a band attachment is solved in an very aesthetically pleasing and original way. Despite the large care with manufacturing the movement, the band attachment is done rather simply. A 27.07 diameter movement has been put in the case. Early Omega models did not have seconds.

The inscription had to be done in or after 1891 when the company name was changed from "Louis Brandt & Fils" to "Louis Brandt & Freres". The movement is a rather common caliber in bridge layout with an unusual 29.33 diameter. The slide to activate the repetition is at 3 o'clock.

The 'Wandering' Crown

In 1898 the Omega Caliber diameter 42.86 was also manufactured in a smaller version, and soon thereafter the Brandts put the wristwatch in their line of products. Those were not exceptional models or technically spectacular creations, but rather utility articles with a long lifespan. By the turn of the century, Omega started manufacturing wristwatches in small quantities. This is confirmed by a large advertisment in the Leipzig Watchmaker's paper, which introduced Omega's new watch type. The ad described the practical usefulness of wristwatches in extreme situations. A British officer in Canada had bought a dozen Omega wristwatches and taken them to Africa where he tested them during the Boer war, which ended in 1902. He spoke very favorably about the watches.

Since approximately 1902, Omega wristwatches were sold which had the winding crown at 9 o'clock. The reason apparently was not for shock protection, but because more and more wristwatches were worn on the right wrist. The band attachment was with wire lugs. The watch did not have subsidiary seconds.

9

Jewelry watch from 1914. The winding crown is at 9 o'clock.

Interestingly, wristwatches—contrary to lady's pendant watches—were missing the subsidiary seconds, which could have easily visually confirmed the running of the movement. Following the pocket watch designs, wristwatches were in hinged cases with cuvettes.

Around 1902 the winding crown moved, for inexplicable reasons, to the left side of the watch case, at 9 o'clock. This apparently was not done for protection of the movement from shock, but in order to promote the wearing on the right wrist. This was by no means a short term phenomenon—two large watches in the Omega catalogue of 1906 have the crown on the left. As late as 1914, among the twenty-five Omega watches featured in the catalogue for the national exhibition in Bern, there is still a watch with the winding crown at 9 o'clock.

A large part of the production of these early wristwatches was made for the Anglo-Saxon market. The English watchmaker Edwin J. Volkes, located in Bath, had watches without the Omega logo or with his own trade name. The largest seller of Omega watches (up to the Second World War) was the Paris department store Kirby, Beard & Co, whose logo was on the lower half of the dial, while the Omega logo was in the upper half. Very early models had a coin edged crown, which was slightly removed from the case and sitting in a small tube to hinder water and dirt from entering the movement.

Military watches from World War I again had the crown at its original position at 3 o'clock. Some of those watches had a protective pierced stainless steel cover, and some had radon luminescent numerals and hands to allow for use in the dark.

Simple, yet strong and curved lugs soldered to the case, allowed the leather strap to be attached to the case. Those watches did have subsidiary seconds. Towards the end of the war these watches were delivered to the British Air Force and an American Expedition Corps (Signed: Signal Corps USA).

Omega watches from later years, which do not have the winding and setting crowns at 3 o'clock, do exist. There is a gentlemen's watch, for example, with modern lugs and a low numbered movement (below 5,000,000), whose dial is turned in a way which makes the crown and 12 o'clock actually at 2 o'clock. While the aviator's watches of the 1930s had large ball-

Wristwatches featured in the main catalogue of 1906.

Military watches from World War I.

form crowns in order to be turned while wearing gloves, the trend for jewelry watches tended toward placing the crown on the back case. Luxury watches sometimes had the crown under the bezel or sunken in the case. Recessed crowns set within protective saddles were a development of the 1980s. Omega has numerous wristwatches with such crowns.

Rare Omega model with turned dial.

Large aviator's watch from 1934 with unusual winding crown. The revolving crystal bezel is calibrated to keep track of and read the flying time. The Staybrite steel case with strong lugs and leather strap allows the watch to be worn over the sleeve or on the leg. Black contrast dial with luminescent numerals and radon hands allowed for good night use. The watch was made with a flat pocket watch caliber 35,5 S T1. 35,5 being the movement diameter, S standing for subsidiary seconds and T1 for 1. Modification of the original 1918 Caliber.

Diver's watch Omega Seamaster Professional from the 1980s, water-resistant to a depth of 200 meters. One of the models with protected recessed winding and setting crown.

Omega Constellation from the 1980s with protected recessed crown. The gentleman's model has two subsidiary dials for the date and day of the week.

Pilot's Wristwatches

Even though the cockpit is equipped with all the necessary time measuring instruments, pilots wanted their own personal watches, to be worn either over the sleeves of their suits or around the upper thigh. Therefore, in addition to board instruments, Omega in Biel developed special watches for aviators.

In the first half of the 1930s a large wristwatch was custom made for this purpose. It had a large black dial, large numerals with a radon layer, and a revolving bezel to mark takeoff and flight time. Strong, long lugs attached the time instrument to the leather band. The watch was equipped with Caliber 35,5 S T1. The 16 ligne movement with 15 jewels had been developed in 1918, originally for thin pocket watches. (Ill. page 14)

By 1930 the Caliber 35.5 was replaced by the new and larger Caliber 37.6. The Staybrite Steel case had a diameter of 44.5 mm and was strongly reminiscent of the previous model from the first half of the 1930s, except that the lugs were much shorter.

A second model, Reference CK 2042, was available in 1940 with a massive steel case, strong, curved lugs with drill holes for the pins, and a small, flat winding crown. Despite the small Caliber 26.5, the overall diameter of the watch was 40.5 mm. Both models continued having black dials.

Besides the revolving bezel with calibration and the subsidiary seconds, these watches did not have any additional complications. Compare this to the Omega Flightmaster from 1969, which had two crowns at 8 and 10 o'clock which were needed to set the outer chapter ring calibrated for 60 minutes and the second time zone, given by a second hour hand. As additional functions, the watch had a chronograph mechanism without a column wheel, a 12 hour register, and either subsidiary seconds (Caliber 911) or a subsidiary dial for 24 hours (Caliber 910). The ebauches were made by Lemania. The case had a cover with bracelet attachment.

CK 2000 37,6 acier inoxydable staybrite . . . Fr. 78.—

∅ 44,5 mm   Ouv. lun. 30 mm   Ecart. anses 22 mm
                   Cadran 755 ox. radium Fr. 5.—

CK 2042 26,5 acier inoxydable staybrite   Fr. 85.—

∅ 40,5 mm   Ouv. lun. 28 mm   Ecart. cornes 24 mm
                   Cadran 755 ox. radium Fr. 5.—

*The aviator's wristwatch CK 2000 and CK 2042 in the
Omega Main Catalogue from 1940.*

Omega Flightmaster Caliber 910 (with 24 hour dial), 1969. The watch has a second hour hand for dual time zone.

Wristwatch with Chronograph

The chronograph wristwatch was made in 1909 and was always considered special. While the early models allowed for the measuring of time intervals up to 30 minutes, the early Omega chronograph, available since 1912, only had a 15 minute register. The center chronograph hand could be started, stopped, and returned to zero with a push button at 6 o'clock. The subsidiary dial for the register was at 3 o'clock and the constant seconds dial at 9 o'clock. It was a large watch equipped with a pocket watch movement.

For the next 20 years Omega continued producing the large chronograph watches. Even a new model, made from 1928 on, did not comply with the need for a smaller wristwatch—its diameter was 39mm. This Caliber was used for car and cockpit watches, as well as pocket and wristwatches. However, the chronograph wristwatch did receive nicer and more attractive new lugs compared to the earlier ones, which were soldered.

Only in the 1930s when the most important time for chronograph wristwatches started, did Omega also focus on its chronograph program. From 1932 on the Ebauches for chronographs were purchased from Lemania in L'Orient on Lake Geneva. They were classified as 28.9 CHRO and 33.3 CHRO. The mechanism was activated by one push button.

In 1933 the chronograph division took prominence in the company history—Omega had beaten the large competition and secured the account of official supplier to the Italian Air Force.

Very early chronograph wristwatch. The Omega model had a 15 minute register.

Italian Air Force Secretary Italo Balbo, known for his sensational formation flights over long distances, among them one to South America in 1930, ordered Omega Chronographs for the crew of his water airplanes in preparation for an aviation race from Rome to New York. A total of 25 machines under the Italian flag were produced for this trans-Atlantic adventure.

In 1935 the Caliber 33,3 CHRO received its first modification, followed by four more by 1959. The new chronograph generation had two push buttons, allowing for additional timing options. This meant that the timing process could be stopped and continued at a later point.

In 1941 Omega added to its program the small Caliber 27 CHRO,

Small Omega chronograph in beautiful case with an attractive dial. This model was made circa 1935.

which was offered with a 30 minute or 12 hour register. It replaced Caliber 28,9 CHRO. Chronographs with Caliber 28.9 CHRO were no longer offered in the main catalogue of 1946 (G 580). The prospective customer could choose between 30 different models, with case sizes varying from 32.50 to 38 mm diameter. The two water-resistant-type models could be distinguished by their round push buttons, while the eight watches with Caliber 33,3 CHRO had elliptic push buttons. The most expensive piece was a 27 CHRO with a 12 hour register in an 18k gold case, which cost 634 Swiss Franks, while the sale watch in a Steel case cost 292 Swiss Franks.

In 1949 Caliber 27 CHRO, after having been modified in 1946, received the new name 321, the model with minute register Caliber Nr. 320 respectively. The larger Caliber 33,3 CHRO was renamed 170 (with minute register) and 171 (with 12 hour register). The added

21

Two chronograph wristwatches by Omega

The model on the left is in a steel case, circa 1940, with tachometer and telemeter calibration on a pink and grey dial. The tachometer calibration is in spiral form starting in the center of the dial. Tachometer calibrations are used to time the speed of an object at a distance of 1000 meters. The reading is done as follows: If the minute register after the timing is still at zero, the reading is done on the outer chapter ring (speeds between 1000 and 60 kilometers per hour), if the minute register is at 2 minutes, the speed is read on the inner ring of the spiral (speeds between 30 and 20 kilometers per hour). The illustrated watch shows a speed of 35 kilometers per hour. The sound telemeter calibration on the outer ring is calibrated for kilometers and is based on the speed of sound in the atmosphere (approximately 340 meters per second). The chronograph hand of the illustrated watch is at 13, meaning, for example, that the thunderstorm is in 13 kilometers distance, because between the lightning and the thunder, the hand has moved from 0 to 13. The watch has two push buttons, the second one, which returns the chronograph hand to zero is set in the winding crown. The model has a Cal 33,3 Chro movement. The sportiv leather band has a clasp.

The model on the right is from the 1920s. The beautiful hinged case is in yellow and gold and houses a Caliber 30 Chro antimagnetic movement. The tachometer scale is calibrated for a distance of 1000 miles and is 1/5 of a second scale. The subsidiary dial with the 30 minute register is located at 3 o'clock and is slightly larger than the one for constant seconds located at 9 o'clock. The wide push chronograph button is located at 6 o'clock, where it can be easily acitvated with the thumb.

23

construction changes and improvements were an Incabloc shock resistance, and an increase of jewelling, anti-magnetism, etc. All calibers have a chronograph mechanism with a switching wheel.

The height of the watch could thus be reduced continuously. While the 39 CHRO had a height of 7.15 mm, the height of the 33,3 CHRO was lowered to 6.50 mm and the 27 CHRO with minute register to 5.57 mm. The 27 CHRO with an hour register requires a height of 6.74 mm.

From 1957 on Omega sold the Wristwatch Chronograph Caliber 321 under the name Speedmaster. This watch with manual winding was also sold in the United States, where, unknown to the company, the United States Aerospace Agency NASA had purchased the watch along with nine other models and makers, and submitted it to rather strict tests. The best watch was supposed to be the one used by the astronauts. Omega achieved unbelievable results and in 1965 was chosen as the astronaut watch. From 1966 on it received the surname Professional. From then on every astronaut was wearing this Swiss precision watch into space. The Speedmaster Professional was part of 50 space missions, including 6 moon landings from the Gemini space flights, the Apollo and Skylab programs, to the recent space shuttle flights. Neil Armstrong was wearing a Seamaster Professional on his wrist when he was the first man to set foot on the moon on July 21, 1969. A Seamaster was also part of a Apollo-Sojous Meeting between Americans and Russians in space on July 17, 1975. On April 17, 1970, when the Apollo XIII mission had to be cancelled under dramatic circumstances, the Omega Speedmaster Professional allowed the crew a safe return to earth. After the failure of all board instruments, the astronauts had to use their chronograph wristwatches to calculate the exact moment of ignition for their engines to return to earth. The Bieler company has received the Snoopy Award, the highest award given by NASA.

pages 25 and 26:
Gold Omega Chronograph Wristwatches with registers for 30 minutes and 12 hours from before and after 1949. The more recent model from the 1950s has the new name Nr. 321 for the old Caliber 27 CHRO inscribed on the balance block.

CHRONOGRAPHES 27 CHRO

17 rubis

Antimagnétiques et pare-chocs

avec compteur d'heures

ÉTANCHE

CK 2451 27 chro. C 12 acier inoxydable, glace
incassable .   .   .   . Fr. 382. —

⌀ 35 mm, ouv. lun. 29,5 mm, cuir 18 mm, cadran 2206 ardoise radium

pages 27 to 31:
The chronograph wristwatches
offered by Omega in 1946

27

Avec compteur d'heures

CK 2277 27 chro. C 12 acier inoxydable . . Fr. 334. —
OJ 2277 27 chro. C 12 or 14 ct. . . . . „ 550. —
OT 2277 27 chro. C 12 or 18 ct. . . . . „ 613. —

⌀ 32,5 mm, ouv. lun. 28 mm, cuir 18 mm, cadran 2208 tachymètre

Sans compteur d'heures

CK 2276 27 chro. acier inoxydable . . . . Fr. 292. —
OJ 2276 27 chro. or 14 ct. . . . . . . „ 508. —
OT 2276 27 chro. or 18 ct. . . . . . . „ 565. —

⌀ 32,5 mm, ouv. lun. 28 mm, cuir 18 mm

CK 2381 27 chro. acier inoxydable . . . . Fr. 307. —
DB 2381 27 chro. plaqué or 20 microns . . . „ 356. —
DP 2381 27 chro. plaqué or 40 microns . . . „ 386. —

⌀ 37,5 mm, ouv. lun. 32 mm, cadran 2213 points relief

Avec compteur d'heures

CK 2279 27 chro. C 12 acier inoxydable . . Fr. 341. —
OJ 2279 27 chro. C 12 or 14 ct. . . . . ,, 565. —
OT 2279 27 chro. C 12 or 18 ct. . . . . ,, 634. —

⌀ 35 mm, ouv. lun. 30 mm, cuir 18 mm, cadran 2206

Sans compteur d'heures

CK 2278 27 chro. acier inoxydable . . . . Fr. 299. —
OJ 2278 27 chro. or 14 ct. . . . . . . ,, 523. —
OT 2278 27 chro. or 18 ct. . . . . . . ,, 586. —

⌀ 35 mm, ouv. lun. 30 mm, cuir 18 mm, cadran 2206

OJ 2380 27 chro. or 14 ct. . . . . . . Fr. 540. —
OT 2380 27 chro. or 18 ct. . . . . . . ,, 607. —

⌀ 35 mm, ouv. lun. 30 mm, cadran 7017 heures or rivées
Fr. 66. —

29

33,3 CHRO

17 rubis

Antimagnétiques et pare-chocs

STANDARD

CK 987 33,3 chro.   acier inoxydable .   .   .   .   Fr. 272. —
OJ 987 33,3 chro.   or 14 ct. .   .   .   .   .   „ 464. —
OT 987 33,3 chro.   or 18 ct. .   .   .   .   .   „ 534. —

⌀ 37,7 mm, ouv. lun. 32 mm, cuir 20 mm
Cadran 2213 pt. relief tachymètre, pulsomètre

ÉTANCHE

CK 2077 33,3 chro.   acier inox., glace incassable .   Fr. 319. —

⌀ 38 mm, ouv. lun. 32 mm, cuir 20 mm, cadran oxydé 2039 radium

CK 2393 33,3 chro. acier inoxydable . . . . Fr. 272.—
OJ 2393 33,3 chro. or 14 ct. . . . . . „ 464.—
OT 2393 33,3 chro. or 18 ct. . . . . . „ 534.—

⌀ 37,7 mm, ouv. lun. 34 mm, cuir 20 mm
cadran 2220, chiffres polis, tachymètre

CK 2404 33,3 chro. acier inoxydable . . . . Fr. 275.—
DB 2404 33,3 chro. plaqué or 20 microns . . „ 329.—
DP 2404 33,3 chro. plaqué or 40 microns . . „ 359.—
OJ 2404 33,3 chro. or 14 ct. . . . . . „ 517.—
OT 2404 33,3 chro. or 18 ct. . . . . . „ 605.—

⌀ 37,5 mm, ouv. lun. 34 mm, cadran 2215

31

From 1968 on, the "Moon watch" was no longer offered with Caliber 321, but with Caliber 861 instead. It had a chronograph mechanism with a newly developed, very sturdy control lever mechanism in the Landeron style (without column wheel). The movement had a glucydur balance, eccentric fine regulation, and had a higher frequency of oscillations (21.600 half oscillations as compared to 18,000 before).

In 1968, a Speedmaster Professional in a stainless steel case with screwed back was part of the Plaisted-North Pole expedition. The watch was a reliable companion for the expedition members for 44 days, despite temperatures as low as minus 52 degrees Celsius. In 1978 NASA launched a competition for a new astronaut watch to be used in its space shuttle program to be started in 1981. No less than 30 makers competed. Omega submitted its already tested Mechanical chronograph with manual wind and won the competition—the Omega Speedmaster Professional continued to be the Space watch. The tested watches had to sustain temperature changes from minus 18 degrees to plus 93 degrees Celsius. They were kept in vacuum chambers for days and dropped to 15 meters under water. They were accelerated to 16 times the gravitation of the earth, and exposed to vibrations of 5 to 2000 Hertz and shocks of 40 times the gravitation. The watches were also tested for corrosion resistance in a pure oxygen environment.

This renewed "flight qualified" honor was taken as an incentive to launch a limited numbered edition of this exceptional watch in an 18k gold case with the inscription "Apollo XI—first watch worn on the moon." A sapphire crystal back allowed for the viewing of the movement. When the renowned Munich watch dealer Huber celebrated its 125th anniversary, the only chronograph wristwatch offered in the anniversary catalogue was the Omega Speedmaster Professional of a limited edition. The watch was offered for DM 19.500 in yellow gold and DM 23.000 in white gold.

Today this "Himmelsstuermer" ("Sky Climber") is sold in the classic original version in a stainless steel case with screwed back, with a Speedmaster Medallion and steel bracelet or in a special edition with a sapphire back case, which allows one to see the hand polished and adjusted movement underneath. Both watches have the back case engraved with the words: "The first watch worn on the moon."

In 1992 the Bieler company celebrated the 50th anniversary of its Caliber 27 CHRO C12 with three numbered series of Gold Speedmaster Professionals. Nine hundred and ninety-nine pieces were made of the standard Caliber 863 version, 250 pieces of the chronometer version with sapphire back crystal and Caliber 864, and 50 pieces in a skeletonized version with Caliber 867, made entirely by hand.

In honor of the 125th anniversary of Omega in 1973, the watch was offered in a special model with automatic movement, a subsidiary 24 hour dial and aperture for date. Instead of the small 30 minute register, the watch had a central 60 minute register, with the 12 hour

Speedmaster 125, the anniversary watch from 1973. Central minute register, subsidiary dial for hour register at 6 o'clock. Chronometer watch with date and 24 hour subsidiary dial. Movement driven by Automatic Caliber 1041.

register moving a one hour increment after each full turn of the minute register. The Speedmaster 125 was a Chronometer watch and had a Caliber 1040 movement from 1970. Due to its qualification as a chronometer it was named No. 1041. It had a diameter of 31 mm. Subsequently a standard version of this model was put on the market as Speedmaster Professional Mark III Automatic.

Following models were Caliber 1045 with Date and Day of the Week (1975) and Caliber 1140. Latter has the ETA ebauche 2892-2. It is, in addition to the omission of the calendar, adapted accordingly. The back side of the movement is needed for the rotor winding, and the chronograph mechanism which is mounted on an additional plate had to be set under the dial. This does have the disadvantage, how-

33

Speedmaster Chronograph with Automatic Caliber. Model from the late 1980s.

Page 35: Special model Speedmaster Professional in Luxury Chronometer Edition, 1992, in numbered and limited edition (250 pieces). The model is sold as reference number PIC 3194.50.00. It is equipped with Caliber 864 (manual winding, parallel gearing).

Page 36: The skeletonized version of the Speedmaster Professional Chronometer from 1992. The edition size was 50. The skillful work of the engraver can be viewed through the skeletonized dial plate and the sapphire back case. Reference number PIC 3696.50.81

Page 37: The titanium chronograph with Automatic Caliber and Chronometer certificate (Reference number PIC 5890.40.00). From 1993.

Speedmaster Automatic (Self-Winding Chronograph with date in 18k gold case with sapphire crystal).

ever, that for testing or repairing the chronograph, the chronograph gears are not easily accessible, as is the case with classic movement arrangements where the watch gears remain untouched. This model also does not have a screwed back case, however, the bracelet does have the reliable Omega clasp. This model is also sold with leather strap.

The current Speedmaster series offers the Automatic Chronograph with registers for 30 minutes and 12 hours and date, but without day of the week, 24 hour dial, and 60 minute central registers. The same model is offered in a special series in a chronometer version in either a titanium case or in gold with leather strap, but without a rating certificate.

In 1967 Omega added to its chronograph program a stopwatch for the wrist which was made for younger wearers.

Caliber 865 had been developed for the Chronostop, which had a simple yet very sturdy chronograph mechanism without column wheel and a single push button. Later this model was offered with a date (Caliber 920). Both Calibers had manual winding and an eccenter fine regulation. The ebauche was supplied by Lemania. The Caliber 930 chronograph, on the market from 1969, was a two button chronograph, with a 30 minute register at 3 o'clock, subsidiary seconds, and aperture for date at 9 o'clock. The movement had 17 jewels and could be adjusted.

Screw Bezel,
Swing Ring:

Silver . . . . £ 4.17.6
9 ct. solid Gold £ 7. 5.0

Screw Bezel,
Swing Ring, open.

Water-protected Omega from 1931.

39

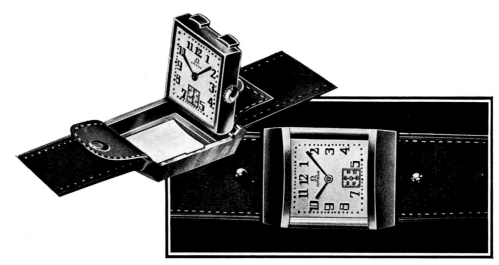

Omega in its fight against sweat, water drops, rain, humidity and submersion in water: special case with hinged double case from the 1930s.

Advertisment for a water-resistant watch, such as Reference 3532.

The water-resistant
Marine model from the
1930s. The impregranted
leather strap has an
adjustable deployant
buckle. See also page 45.

Pages 42 to 45:
Omega's water-resistant models from 1940.   4 1

CK 2077 33,3 chro.   acier inoxydable staybrite
glace incassable . . . Fr. 205.—

Cad. No. 985 Télémètre, Tachymètre, Pulsomètre   Majoration Fr. 13.—
⌀ 38 mm   Ouv. lun. 32 mm   Ecart. cornes 20 mm

CK 2076 28,9 chro.   acier inoxydable staybrite
glace incassable . . . Fr. 245.—

Cad. No. 805 Télémètre, Tachymètre, Majoration Fr. 11.—
⌀ 32,5 mm   Ouv. lun. 27,5 mm   Ecart. cornes 18 mm

Modèle „Marine" pour dames

CK 3673 R 13,5 acier inoxydable staybrite
                glace incassable . . . Fr. 100.—
OJ 3673 R 13,5 or 14 ct. glace saphir . . Fr. 230.—
OT 3673 R 13,5 or 18 ct. glace saphir . . Fr. 270.—

Ouv. lun. 16 × 12 mm   Ecart. cornes 12,5 mm   Cadran 864 ox.

Modèle „Naiad" pour dames

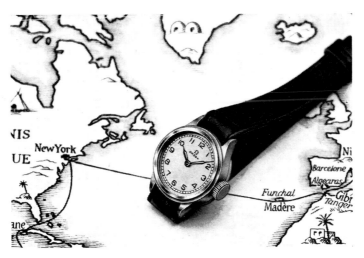

CK 2086 R 11,5 acier inoxydable staybrite
                glace incassable . . . . Fr. 122.—

Ø 21 mm   Ouv. lun. 17 mm   Ecart. cornes 10 mm   Cadran 896

43

Marine Standard CK 3635 T 17   acier inoxydable staybrite
avec glace incassable . . Fr. 100.—

Ouv. lun. 21,5×15,5 mm   Ecart. cornes 17,5 mm   Cadran 445

Marine de Luxe CK 3637 T 17   acier inoxydable staybrite
avec glace saphir . . . Fr. 115.—
OJ 3637 T 17   or 14 ct. avec glace saphir . Fr. 355.—
OT 3637 T 17   or 18 ct. avec glace saphir . Fr. 445.—

Ouv. lun. 19,5×15,5 mm   Ecart. cornes 17,5 mm

Boîtier en acier inoxydable staybrite - Glace incassable ou
saphir sertie dans la lunette - Fermeture interchangeable -
Breveté, sans vis, ni charnières - Montres garanties
imperméables sous 2 atmosphères de pression

CK 679 19,4 acier inoxydable staybrite . Fr. 125.—
brac. cuir avec fermoir spécial en acier . Fr. 11.—

Ouv. lun. 15,8 × 15,8 mm   Ecart. anses 18 mm

OJ 680 19,4 or 14 ct. . . . . Fr. 443.—
OT 680 19,4 or 18 ct. . . . . Fr. 545.—
brac. cuir avec fermoir spécial en plaqué or Fr. 17.—

Ouv. lun. 15,8 × 15,8 mm Ecart. anses 18 mm Cadran 654 Fr. 2.—

Glace saphir - Couronne de remontoir protégée
Cuir inaltérable - Fermoir ajustable

Pages 46 and 47:
The water-resistant model Marine Standard, which
was sold with Caliber T17. See also page 44.

45

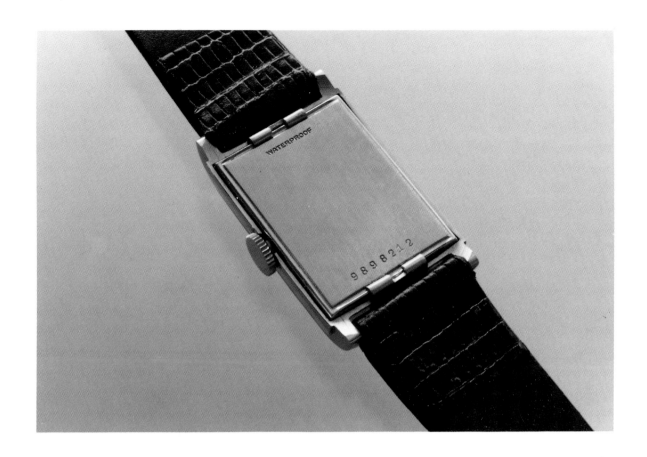

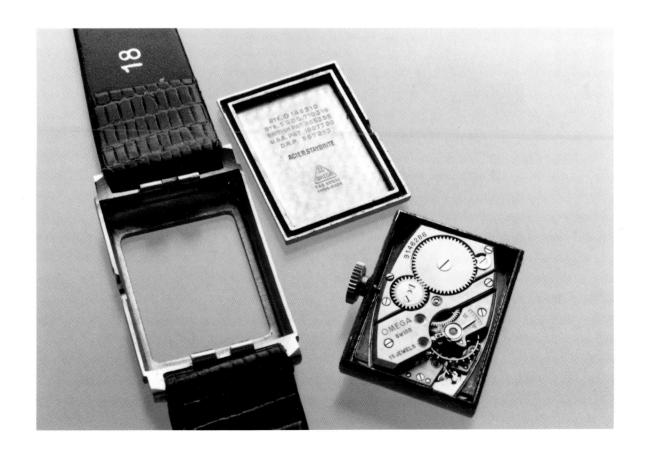

Omega Wristwatch for the Royal Airforce from 1953: water-resistant Staybrite case with double back case, leaded seals, soldered lugs, radon numerals and hands, Caliber 281 (30 SC T2) with sweep seconds and special regulator.

From Water-protected to Water-tight

The protective case originally consisted of a hinged case with a cuvette, the so-called dust cover. For cost reasons the cuvette became more and more left out. However, when the watch manufacturers realized that a tight case was absolutely necessary, they started to look for a solution to this problem with more or less ambition. For most watchmakers, the path to a water tight case proved to be a long one. Omega took much longer, for example, than Rolex, but from the 1930s on the Bieler company was serious about adding models with special features which would protect the movement from water and humidity to its collection. Yet the expenses were enormous for the poor results achieved. In order to increase the everyday use, Omega used a shell construction with screwed on upper case.

Another model with a rectangular case received a second outer case in order to protect crown and movement from water. The rectangular and curved Reference 3532 in a Staybrite Steel case had a triple protection for the movement. Another solution to the problem was found in the 1930s with the rectangular Marine model, which had a second case sliding over the actual case.

The main catalog G550 from 1940 gives an overview of the Omega's achievements in this area. The Marine model had its own small collection. The Marine model in its sliding double case was offered in steel or gold and was equipped with Caliber 19,4 from 1923. The Marine Standard and Standard Deluxe—rectangular gentlemen's watches—both had the tonneau Caliber T 17 from 1934. Interest-

The first Seamaster series from 1948. Both models in stainless steel case. The model on the right was the standard edition, the one on the left with chronometer certificate. The movement was Automatic Caliber 28.10. The silvered matte dial had applied hour and triangular numerals.

ingly the Marine model also came in a lady's size. It was also rectangular and contained the new Form Caliber R 13.5. Another water-resistant lady's watch was offered in a rectangular stainless steel case (diameter 21 mm). It was equipped with the Form Caliber R 11,5 from 1936. The model was named Naiad and sold for 122 Swiss Francs.

Two different Chronograph wristwatches were offered in water-resistant-cases. The Caliber 28.9 CHRO had a diameter of 32.50 mm, the one with Caliber 33.3 CHRO had a diameter of 38 mm. Both had two push buttons and a 30 minute register.

49

The Diver's Watch—Seamaster

The two World Wars brought about a high demand for sturdy military watches. By 1945 Omega had delivered more than 110,000 wristwatches in water-resistant-style cases, with black dials, large numerals, and sweep or subsidiary seconds to the Royal Airforce in England.

This aviator's watch was the forerunner to the diver's watch Seamaster. Its birthyear was 1948. From the first edition on it was considered a noticeable watch, that was purposely equipped with a self-winding movement; a further protection of the crown sealing, since the watch did not have to be wound every day. From 1956 on the self-winding Caliber 501 (with 19 jewels, and swan's neck regulation) was used primarily. The wristwatch with the seahorse trademark made itself known very fast. Ultimately, the guaranteed diving depth was not sufficient. In order to meet modern diver's demands and to keep up with the competition, the water tightness had to be improved. Whereas during the 1950s a depth of 30 to 60 meters was sufficient, in 1981 the top models were absolutely watertight and usable down to a depth of 120 meters.

In 1988 the Model 200 was added to the collection. It also received the surname Professional. The Professional was offered in three sizes and in stainless steel or stainless steel and 18k gold. Its main characteristics were a luminescent dial, luminescent hands, scratch proof sapphire crystal, screw-down crown with side guards, fluted revolving bezel calibrated for 60 minutes, case with screwed seal ring, and bracelet with a double safety buckle. The movement was either Quartz Caliber 1438 or rotor Caliber 1111. The Seamaster was also available as jewelry watch and as chronograph.

Heavy Gold Seamaster with integrated bracelet.

Chronometer wristwatch Seamaster in rare case.

Seamaster Caliber 503 (special advance construction, 20 jewels, sweep seconds, date).

Seamaster chronograph wristwatch with 12 hour register (Caliber 321) in heavy yellow gold case from 1958. The retail price of this watch with gold hands was 765 Swiss Francs.

Seamaster Professional Chrono Diver

In 1993, Omega surprised the industry and watch collectors alike with a new edition of its entire Seamaster collection. The highlight of this new series, and of the entire Seamaster series since 1948, is the Seamaster Chronometer Diver with Chronograph, water-tight to depth of 300 meters.

This diver's chronograph with its new overall look is a small wonder; it is the beginning of a new era for Omega. The beautifully designed watch needs to be examined in further detail. The watch is offered for underwater use in either a titanium case (DM 4,500) with titanium and a 18k gold band, bezel, and crown (DM 5,900); or in titanium, gold, and tantal. (DD 6,900). The unmistakable case with its distinct bezel and the shaped guard for crown protection has a diameter of 42.50 mm and a height of 16.40 mm. The back case has a wavy surface for slip protection along with the traditional Seahorse emblem and SEAMASTER inscription. The two case parts are sealed with a ring, which is bedded in a groove. The movement has an additional movement cover. The crown is screwed on and further sealed with a ring similar to a pressure seal. The patented, non-screw chronograph buttons withstand a pressure of 30 atmospheres and can be exchanged without opening the case. A helium pressure valve with screwed crown at 10 o'clock allows for the pressurizing and depressurizing of the watch when ascending from deep waters to avoid an explosion. The bezel revolves counter-clockwise and the sapphire crystal is non-reflective. The dark matte dial has a wavy surface, and the large hands and tritium numerals make a misreading of time virtually impossible. All Chronograph hands are red and nicely contrast the dark background. The aperture for date is as usual at 3 o'clock. The titanium bracelet with curved links and surfaces conforms to the wrist. The heart of the watch comes from ETA, with several improvements to the otherwise simple Caliber 7750. It is rhodium plated, damascent and of chronometer quality. The automatic movement has a central rotor. Its Omega number is 1154.

The Diver 300 m is a total novelty, as it is the first diver's chronograph usable up to a depth of 300 meters. Other diver's watches may be usable much further down, yet none of them has a chronograph feature.

One of the most sturdy diver's watches ever built by Omega was the Seamster 600, with its very unique exterior. The angular monocoque case with circular bezel had a crown entirely set into the case at 9 o'clock; in addition to being set into the case it had a guard on top of it which could be locked to the case. The deeply fluted bezel could be rotated in either direction after unlocking it by pushing a red button. The underwater researcher Jacques Cousteau was wearing this exact model during his Janus expedition in the Gold of Jaccio, close to Corsica. The expedition included two dives per day in a diving bell, with each dive lasting approximately two hours.

The leading model Seamaster Professional Chronometer ChronoDiver 300 with Date in a Titanium case with gold details.

The Ploprof from 1970. Its official name was Seamster 600, because it would withstand a pressure of 60 atmospheres.

57

The first chronometer wristwatch by Omega, which was an award winner at
timing contests at observatories from 1940 on. The watch is in a stainless steel
case with a gold cover and was equipped with the new 30mm precision caliber
and sweep second. The two-tone silvered dial has spider lines and 1/5 of a
second calibration for precise readings. The watch was obviously not a rated
chronometer with an official rating certification, but a so-called observatory
chronometer.

Timing Contests and Omega

Omega was a successful and regular participant in timing contests at the observatories in Geneva, Neuenburg, and the National Physical Institute in Kew-Teddington. After developing the Caliber 30 mm, whose qualities raised high hopes, Omega started sending wristwatches to competitions every year from 1940 on. In its first year the company set a new precision record in England. It was the first time that such a small movement reached 90.50 out of 100 possible points. Omega even beat Rolex, the specialist in chronometer wristwatches. During the contest of 1946 the previous result was even further improved to 92.7 points. In all, this contest lasted for a period of 44 days.

Only in 1945 was this watch allowed to be submitted to the Geneva Observatory. A New Category D had to be created for it. The top rating was 1000 points. In this contest Omega reached the highest mark in its first year, however, participation was fairly limited. Of the ten watches submitted, only eight could actually fulfil the requirements and were tested. The following results were reached:

1. Omega, movement # 9378500, Regleur A. Jaccard   770 points
2. Patek Philippe   720 points
3. Patek Philippe   688 points
4. Patek Philippe   666 points
5. Rolex   639 points
6. Patek Philippe   599 points
7. Omega, movement # 9378513, Regleur A. Jaccard   569 points
8. Patek Philippe   516 points

Since Category B was almost exclusively entered by Tourbillon watches, Omega started its production with one dozen Tourbillon wristwatches, mainly in order to strengthen its lead in category D. When the Biel company sent the first tourbillons to Geneva and Neuenburg to compete, this was considered a sensation, because except for Lip in France, nobody so far was able to show such a product. Within one year Patek Philippe also sent a tourbillon. Initially, they did not fulfil the expectations that had been set. The results published by the Geneva Observatory in 1947 show that the regulars had worked hard in the meantime. The master piece once again came from Alfred Jaccard, who had worked

Omega wristwatch with Tourbillon, movement number 1059542. This was probably the last one of the 12 tourbillon in a special edition series. The lowest known number is 1059533, the highest 1059544. This watch was developed by the tourbillon specialist and director of the watchmaker school in Le Sentier, Marcel Vuilleumier, and built by Jean-Pierre Matthey-Claudet. It was regulated by Alfred Jaccard for participation at observatory competitions in Geneva and Nuenburg, however its results were somewhat disappointing initially, only in its fourth year (1950) was it the top ranking watch for the year. After that the 30mm movement with special regulation was ahead once again.

Chronometer wristwatch Caliber 30 T2 5C RG. The movement has between 15 and 17 jewels, the illustrated watch has 16. The movement is without shock resistance. Typical for this movement is the large balance wheel with screws, the fine regulation is a special edition of the 30 mm Caliber. Movement with manual winding in 18k gold case. The caliber name is written on the balance wheel.

for Omega since 1929: two of his Caliber 30 mm watches achieved the two highest results (834,9 and 832 points). The tourbillon wristwatch movement # 10595933 made 736 points and the 6th rank. 1948 was not such a good year: Patek Philippe pushed the Omega tourbillons No. 34 and 36 with 834 and 806 points to ranks two and four. However, the Patek Philippe tourbillon was also in the losing group with 738 points.

In 1949 Omega sent four tourbillons (No. 33, 34, 35, and 37). The results were even worse than the previous year, while Rolex secured a new record with 859 points. Finally, the company's efforts were rewarded in 1950: the tourbillon No. 33 achieved 867.7 points and a Caliber 30 mm 864 points. Patek Philippe's best result was a third place. Tourbillons No. 34, 35, and 38 had also been submitted. In 1951 Omega was successful in defending its leading ranks. However, the victory was not achieved by the tourbillons, but from two Caliber 30mm movements, regulated by Gottlob Ith, who had been working for Omega since 1920. He could increase the record for precision to 870.3 points. The tourbillons numbers 33, 35, and 37 could only receive honorable mention. In 1952 four wristwatches regulated by Ith reached top ranks.

In 1953 Geneva changed its rules for wristwatches. An absolutely flawless wristwatch would be awarded 60 points. The highest results were reached by Omega in 1966 with 56.68 points. Over the course of this year the Geneva Observatory awarded prizes to 57 chronometers by Omega, 34 by Patek Philippe, 30 by Longines and three from other makers. In 1967 this contest was conducted for the last time.

The other Swiss chronometer competition was conducted in Neuenburg, where Longines, Movado, and Zenith watches were submitted on a regular basis. In Neuenburg the flawless watch was awarded zero points. The competition determined the minus points for a watch, which were limited to 16. Omega was also a frontrunner in the Neuenburg competition. In 1949, Omega was winner of the individual watch category for the first time. This was achieved with a rating of 7.2. which by 1966 was close to flawless. During the last contest conducted, the best result was a brilliant 1,97.

These successes found their way into the company chronicles and the press releases sent to the families of those close to Omega, be it either as an important representative in a large town or a little watchmaker in the country.

*The Constellation Grand Luxe with brickwork bracelet. The luxurious
dial made quite a sensation.*

63

The Chronometer Wristwatch Constellation

Chronometer wristwatches became part of the collection of numerous watch manufacturers, especially Rolex and Omega, who jointly widened the markets for such watches. Every chronometer watch is tested at an official rating company and sold with a certificate. Up until the 1950s, Rolex clearly was the market leader for those watches, which changed in 1952 when Omega launched its Constellation Chronometer wristwatch. While Rolex had sold over 135,000 chronometers for the wrist, Omega had managed to sell 8,000 during the same period. Now Omega was filling the gap swiftly. The rating offices were busy testing all those chronometers, and so was production, which had to be increased in order to satisfy demand. The chronometer wristwatch market reached an enormous size.

In 1952 Switzerland issued 42,000 chronometer certificates, 26951 of which were for Rolex watches and 13,954 for Omega. In 1958 the Omega Constellation passed the Rolex Oyster Perpetual Chronometer and eventually left it far behind. By 1963 the relation was 103,041 to 44, 305. In 1969 Omega had 194,580 watches tested and Rolex 179,169—unbelievably high numbers.

The following year, the numbers were reversed again—161,424 Omega chronometers versus 193,790 Rolex chronometers. At this time, Omega had a crisis which had a disastrous effect on its chronometer production (indeed, Omega is not yet fully recovered to this day). Of the 453,799 Swiss watches tested, 440,799 were by Rolex and only 8,823 by Omega. Of those 8,823 watches, 6,614 were actually quartzwatches and 2,209 were mechanical movements. In 1991 the situation was still the same. Of the 666,898 tested chronometers, 626,398 were by Rolex, 9,871 by Tag Heuer and 8,398 mechanical watches by Omega. Mido was number four with 4,251 chronometers.

Chronometer watch Constellation in a heavy gold case. The second generation of this watch had a rotor self-winding mechanism from the mid-1950s on. The actual chronometer movement is Caliber 505 (24 jewels, fine regulator).

A beautiful Constellation from the mid-1950s, when the chronometer was still equipped with the original self-winding Caliber 354 (hammer automatic, swan's neck fine regulation). Slightly modified Constellation dial in luxury edition with the typical applied hourmarkers in gold.

Luxurious constellation model with starry sky dial from 1992.

In 1950, however, the Omega Constellation was a milestone for the company in every sense of the word. It was the flag ship of the company and took Omega to unexpected heights. Just as every Seamaster had its seahorse emblem on the case, a Constellation had the observatory under the night-sky emblem. The dial of the first edition was very luxurious and gave the watch a striking look. The face of a Constellation became an important addition to the different dial styles. Strict lines and sloping planes with applied hour numerals in gold were its trademark. A memorable model for any watch collector is the Grand Luxe with hooded lugs and bracelet composed of brick work links. Omega changed the dial style as well as the circular case for luxury models. By the mid-1960s, a new Constellation with a bow shaped middle section was added to the collection, and in 1970 a shaped case with an integrated bracelet and a dial shaped like a computer or television monitor became available.

The self-winding movement was related to Caliber 28.10; it was not a new development, but had been on the market for ten years, and so therefore was a fully developed product. The chronometer movements were numbered 354 or 355 (with date). The self-winding mechanism with weight allowed for thin case constriction.

The Constellation helped to increase the chronometer production at Omega, and in 1958 Omega surpassed Rolex for the first time: more than 50% of the chronometer certificates were issued to Omega. In 1960 the company proved that it was capable of pairing quality and quantity: 20,000 consecutively produced chronometers were submitted for testing and all of them passed with honorable mention for especially good results. In 1965 100,000 consecutively produced watches received the chronometer certificate, allowing for the chronometer inscription on the dial. In 1967 the one millionth chronometer certificate was issued, and five years later the two millionth.

In 1956 the hammer automatic was replaced by a self-winding mechanism with central rotor Caliber 500. In 1959 Caliber 505, so far used in Constellation watches, was replaced by the chronometer Calibers 551 and 561 (with date). After 1966, the latter had a quick date change. (Caliber 564).

A Constellation with a dial from the late 1960s. Caliber 564 Chronometer Caliber (with date), put on the market in 1966.

The lady's watch with chronometer certificate, Ladymatic, was offered from 1955 on. In 1963 the decision was made to add a lady's Constellation. However, the public interest in such a watch was very limited and could not be further indiced. In 1963 only 37 chronometer certificates were issued, and in 1966 a total of 4,878. The chronometer was equipped with the new Caliber 681, a self-winding movement with 17.50 mm diameter, central rotor winding in both directions, 24 jewels, sweep seconds, and an aperture for the date. The Caliber number 682 was adjusted for five positions and temperatures. A further modification brought a renumbering to 685. The chronometer version without a date had Caliber No. 672.

69

Gentlemen's watches with chronometer certificates were Caliber No. 602, a modification of the original Caliber 600 (diameter 27.90 mm) from 1960, and the slightly smaller Caliber 712 (diameter 25 mm). Same Caliber with date and day of the week with quick date change, was numbered 751 (19,800 half oscillations per hour, 24 jewels, swan's neck fine regulation).

The next Caliber modification took place in the 1970s. The next BULL-Caliber (Bull stands for Bulletin—chronometer certificate) was Caliber 1001, with only 20 jewels compared to 24 before. As a technical distinction, the round movement with 27.90 mm diameter worked on a high frequency of 28,800 half oscillations per hour, had a quick date change, a stoppable second, and off-center minute wheel. However, this movement was far from the quality and sturdiness of the successful Caliber 500 generation.

The automatic Caliber 1111 (diameter 25.60) from the mid eighties was not made by Omega itself, but bought from the ebauche maker ETA, where it was listed as 2892-2. This caliber (without the calendar mechanism) is improved to chronometer quality. Same movement is also used in the mechanical Luxury Constellation from 1992, which has a five tier lacquered dial and a gold set diamond, symbolizing the nocturnal sky. The blue roman numerals are set on the golden bezel. The height of the watch is 7.40mm.

Omega and the Phases of the Moon Indication

When watches with a full calendar and phases of the moon gained popularity, Omega developed such a model. It was one of the novelties of 1947 and carried the name Cosmic, which, however, did not appear on the dial.

Numerous renowned makers were offering a wristwatch with calendar, central date hand, apertures for day of the week and month, and subsidiary dial for constant seconds and phases of the moon. The watches were all fairly similar in a round case. The Cosmic was fairly common in that respect. However, it was also offered in an attractive square case.

70

The mechanical movement had 27 mm diameter, was antimagnetic, shock resistant, and had 17 jewels. The balance made 18,000 half oscillations per hour. Despite the calendar mechanism the movement was fairly flat; it had only 5.25 mm and was similar in design to Caliber 30 mm.

The most important parts of the well-thought out calendar mechanism pivoted on pins and in sockets made of steel and set in holes in the movement plate. The drilled and threaded sockets to hold the screws were set to a certain height in order to allow the lower plane of the screw heads to limit the vertical movement of the freely pivoting parts. The setting grooves for the calendar movement were set in the band of the watch and could easily be set with a pin. In 1949 this movement (27 DL PC AM 17P) was numbered 381.

Omega received orders for this watch from Tiffany & Co. and Tuerler. However, it disappeared from the market only a few years later.

Forty years later, in 1990, Omega did release another watch with a full calendar and phases of the moon. This watch was part of the Speedmaster Classic collection. Since the model had an ETA-Caliber 7751, many functions could be added. Besides the full calendar, phases of the moon, the watch was self-winding, had a chronograph with 30 minute and 12 hour register and a subsidiary 24 hour dial in addition to the large 12 hour dial. Ref 3131.20 was a heavy gold watch with gold bracelet, case diameter 39 mm and 14 mm height.

However this model had the disadvantage that the calendar synchronization between day, day of the week, month, and phases of the moon was valid only for brief periods at a time and needed to be adjusted manually by the wearer.

Quite the opposite, a perpetual calendar wristwatch is a true miracle of micro mechanics. It is pre-programmed for many decades, taking leap years into consideration. Since 1990, Omega has been among the small number of watchmaking companies offering such a miracle of watchmaking in its program.

*The Cosmic by Omega in
different styles.*

The perpetual calendar wristwatch offered in the exclusive Louis Brandt series, availble beginning in April 1992 under Reference PIC 5341.30.13.

The Wristalarm Watch Memomatic

Similar to the calendar wristwatch Cosmic, which came twenty years later, the Memomatic wristwatch remained as a single child with no siblings. The wristalarm neither remained in the collection for any extended period, nor was there ever a succession model developed. The timing of its release was very badly chosen—most manufacturers were already focusing on quartz watches, and Omega itself was restructuring the company at the time.

Wristalarm watches had always been rare, and a self-winding wristalarm was a curiosity. The Memomatic in Dynamic Styling was developed by Omega and had several technical characteristics. The watch got its energy from one spring barrel, which left only a limited amount of power for the alarm. The hammers for the alarm were activated for the duration of one turn of the spring, then the alarm automatically shut off—the automatic needed one full hour in order to regain the necessary energy, assuming that the watch was worn and wound during that period.

The Memomatic from 1969 had an alarm setting precise to the minute. The multi-colored multi zone dial had an outer, firm minute and hour chapter ring for the time and a revolving inner chapter ring for the alarm setting. The winding crown was located at 4 o'clock, and a flat setting crown was located at 2 o'clock. In setting crown position 1, the alarm mechanism was locked, in position 2, the alarm was unlocked and in position 3, the alarm could be set.

The movement had a diameter of 30.80 mm and Caliber number 980. It had 19 jewels, a big sweep seconds hand, and a large aperture for date, with quick date change by a pin in the case band between the winding and setting crown.

The automatic wristalarm by Omega was described in a three-page article in the *German Watchmakers Magazine* in 1972. In the series "The *German Watchmaker Magazine* examines..." the author of the article gives detailed instructions on how to disassemble and assemble its movement (see page 200).

Beautiful and Precious Cases

The Brandt Company was not considered one of the exclusive manufacturers of luxury models, yet it again and again surprised consumers with beautiful and original models. The first line of exclusive lady's watches was done for the exhibition in 1914 in Bern. Two dozen watches of different design were offered, all had a circular movement, due to the lack of a form movement, according to the catalogue.

At the time large rectangular gentlemen's watches were also offered with small circular lever movements. A Curvex from 1915, despite its dimensions of 28 x 47x 9 mm, had only an 11 ligne circular movement.

For travelers by train (women as well as men), watches were offered an option that, in addition to the twelve hour calibration, had a 13 to 24 hours indication in order to facilitate the reading of the train schedules. During the 1920s, the development of wristwatch movements was pursued at Omega in Biel and Geneva, where a second production place had been started, and soon lady's watches, besides the tradition three quarter circular movement, could be set with oval and other form movements. Rectangular gentlemen's watches could be delivered with either a tonneau movement, Caliber 20 F (20x28 mm), from 1928; in 1934 Caliber T 17 (17 x 24.5 x 3.85 mm) followed, which had a running reserve of 60 hours.

The introduction of the oval Caliber 12,3 F in 1922 created many opportunities for designers and case makers. The Art Deco period produced many beautiful and precious lady's watches. Omega was a very ambitious and versatile company, which made many exclusive designs to exhibit at Arts and Crafts shows and therefore attract attention. In 1925, Omega participated at the Expositions des Arts Decoratifs in Paris with a model to be worn on the back of the hand, secured between a ring and a wristband. Omega was awarded a Grand Prize at this exhibition.

In 1914 Omega presented its jewelry wristwatch collection for the national exhibition in Bern on four catalogue pages.

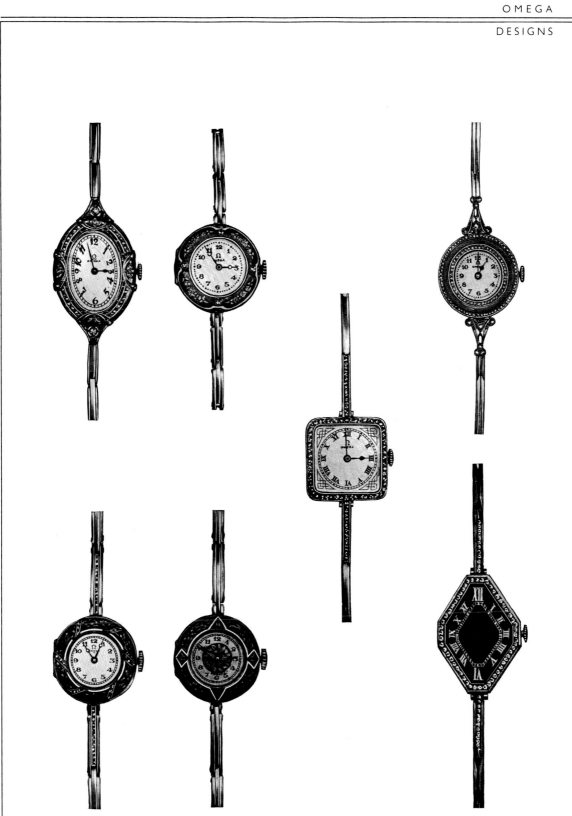

77

78

79

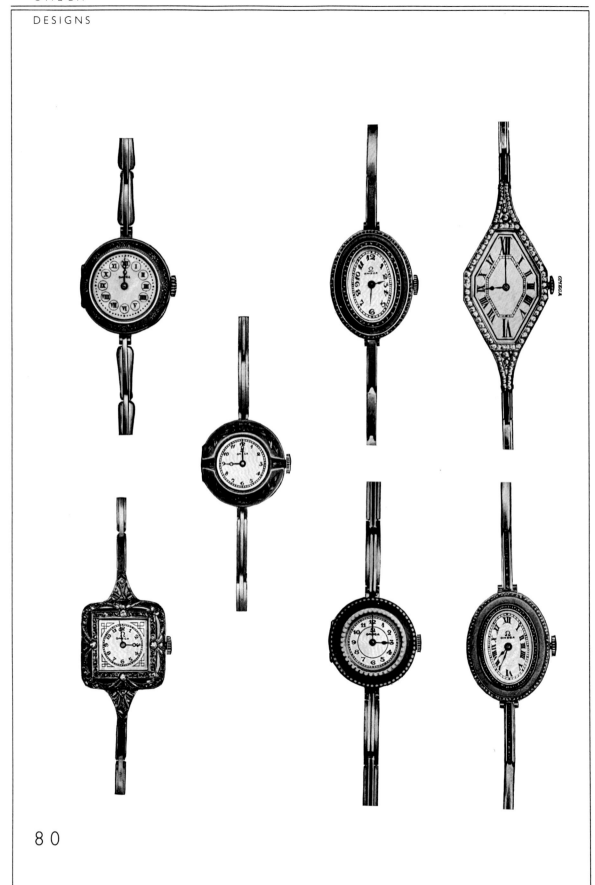

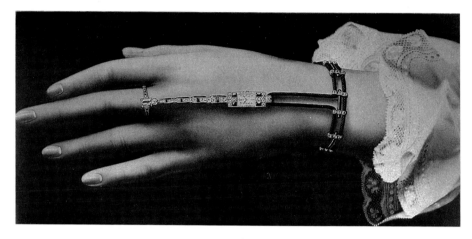

A very original design was this lady's model for the 1925 Exposition des arts Decoratifs in Paris.

Gold and Platinum were the most favored materials for wrist-watches. Omega also made a number of watches in silver until the 1930s. When the wristwatch finally became the market leader and surpassed the pocket watch in popularity, wristwatches were also increasingly put in base metal cases. Some cases appeared to be gold, but actually were made of base metal applied with a thin layer of gold. After the Second World War Omega used a 3/10 mm gold cover, soldered to the top of the case. The first watch with such an indestructible case was Caliber 354.

The snap-on case was replaced with a screwed back case for expensive and water-resistant watches. In addition some watches were decorated with artistic elements: the Seamaster had a seahorse emblem, the Constellation an observatory emblem.

Page 81: expensive lady's watch from 1925. The dial of the platinum bracelet watch, set with eighty-two diamonds, is hidden underneath a lid, set with a large sapphire. The rectangular movement, size 6.55 x 20 x 2.70 mm, is probably not by Omega.

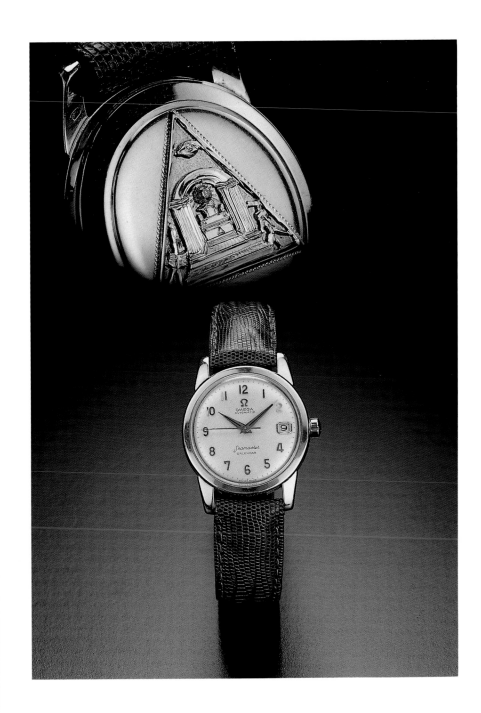

The Free Mason Model of a self-winding Seamaster Caliber 503 from 1957.

From the Habsburg, Feldman Catalogue: A fine gold wristwatch with screwed case and cloisonne dial (a map of Brazil) by Omega from the 1960s.
Photo: Jean-Marc Baumberger

From the Habsburg, Feldman Auction Catalogue: Fine Watches with Cloisonnes dials: the one on the right is an Omega Seamaster with Neptune and Seahorses.
Photo: Jean-Marc Baumberger

Especially beautiful is the bottom of a gold Free Mason watch from the 1950s. Watches with cloisonne enamel dials were custom made for clients from South America and Arabia. Less expensive customized models included, for example, the Golfer's watch, which had a hammered metal dial symbolizing a golf ball.

In memory of the XVI Olympic Summer Games in Melbourne in 1956, a Seamaster Olympic with a Roman XVI on the dial was issued and another model with self-winding movement was also decorated with an Olympic medal, which Omega had received for its outstanding contributions to sports and the time keeping of sporting events at the Olympic summer games in Helsinki in 1952 from the Olympic committee.

When gold bracelet watches became fashionable, Omega initially did not want to comply with the customer's demand for an integrated watch and metal band. When the jewelers started buying gold and metal bracelets from other suppliers, Omega gave in and started offering bracelet watches. It proved to be the right decision. The Omega Constellation DeLuxe with a brick work link bracelet was a very good design. The lady's collection of 1955 was a big commercial success. The Sapphete (with faceted sapphire crystal, simulating a diamond cut) and the circular Ladymatic were very much in demand, especially in gold with gold bracelet. The ladymatic was offered with a gold bracelet similar to the Constellation DeLuxe model. Initially, the ladymatic was sold with a chronometer certificate. In addition, the jewelry watch collection had a large selection of beautiful creations to offer.

For participation in various international competitions, Omega commissioned special watches to be designed by various artists, which frequently received prizes and honorary mention. In 1957, Omega won an Oscar at the Diamond International Award for a gold and diamond set lady's bracelet watch. In 1963 Omega won a total of four Oscars at the same contest and another one in 1964. This third award made Omega a permanent member at the International Diamond Academie in New York, which is the highest organization in the goldsmith trade.

Omega Model with Olympic Emblem, a creation from 1956, when the Winter Olympic Games were held in Cortina d'Ampezzo and the summer games in Melbourne.

In addition, Omega was awarded several prizes at the Prix de la Ville de Geneve. Omega was also very successful at the competition in Baden-Baden, where Omega won over half a dozen Golden Roses as a participant in the jewelry watches category between 1970 and 1978, awarded by an international jury during the German Gem Stone Days.

In 1966 Omega participated in a contest of the Federation Horologer in preparation for the International World Exhibition in Montreal in 1967 and won the top prize for its sports watch Chrono Stop.

Omega began the 1960s with a Monocoque case for its thin Seamaster. It was made from one piece, making the washer obsolete. The shell case allowed the movement only to be reached via the bezel, which had a set in crystal. The water-tight crown consisted of two parts. The crown had to be removed before opening the case. A special tool allowed the crystal to be pressed and removed from the bezel. By slightly turning the movement it could be removed from its shell.

87

Gentleman's wristwatch with gold cushion form case, which was very popular during the 1920s.

Page 89: Ref. 3870, Caliber R17,8, and a square model with calendar at 6 o'clock (Caliber 355).
Pages 90 and 91: Omega model in different form cases from after the Second World War.

Beautiful Omega watch with rare lugs from the 1950s. The gold case holds the automatic movement with rotor winding 471.

In 1966 another modified water-resistant shell type case, the Unicoc, was developed, mainly for the Seamaster model. The removal of the crystal was done by using a so-called Extractor. The crown could easily be removed by using a certain tool, supplied by Omega. The elipse form Omega Dynamic also had this Unicoc case. The watch collector usually is not too happy about this watch, since it is very hard to view the movement. In 1969 the form case Compressor entered the market. It had four wedge-shaped bolt springs in the back case with corresponding slots in the top case.

Pleasing gold gentlemen's watch from the 1940s.

In 1961 Omega suggested a number of special tools for watch-makers in its technical handbook #24. Among those practical items was a massive one size key with massive socket, holding lock, clutch, and lifting arm for opening and closing its water-resistant cases with screwed back. Another tool would allow a professional positioning of the crystal. Another useful tool was a standard tool to measure the water resistance, which neither required a particular water supply nor a pump. A later model used a pressure crank. For adjustment and sizing of silver and lead seal rings, a graduated cone was offered

93

in stainless steel, together with a special attachment to set the ring on the cone. For the automatic watch in stock, Omega offered an automatic winding box.

By the second half of the 1960s, the collection was so extensive that it was hard to keep track. Before the crash of the Swiss watch industry at the beginning of the 1970s, the Omega collection consisted of 2,000 different case styles. Financial problems led Omega to downsize its product line and limit its production of mechanical watches.

As a press release from 1981 shows, the collection by then consisted of the Constellation, Seamaster, and De Ville models with a very limited number of special watches. Most of them had quartz movements. The Constellation line consisted of the following:

> Cristal (with soft geometric line)
> Torsade (with handcrafted design elements, such as gold threads set in bezel and bracelet)
> Marine (with simulated hatch-form bezel and technical design dial)

The majority of the Constellation watches were sold without a Chronometer certificate.

The sports watch Seamaster was offered in two models, the Seamaster 120 (stainless steel, gold plated, or Two-tone, glazed) and Seamaster (water-resistant to a depth of 30 meters, the De Luxe edition in 14k gold and steel or steel with guilloched bezel).

Magnificent gold gentlemen's watch with Caliber 283 movement (indirect center seconds).

The De Ville collection had the following models:

St. Honore (satin finished and polished case, dial with Breguet numerals)
Les Genevoises (with decorated bezel, following Omega designs from the 1930s)
Les Tours de Bras Polonais (gold plated case with Polish-style bracelets with drilled wire mesh)
Tank (rectangular case with decorated bezel and roman numerals)
Les Lagues (with mother of pearl dial or ivory color dial with facetted cabochons for the 3,6,9,12 numerals)
Torsades (two-tone case with screwed bezel and mother of pearl dial) and De Ville design (ultra thin watch in guilloched two-tone case with integrated bracelet)

The exceptional watches mentioned in the press release at the 9th European Watch and Jewelry Fair in Basel were the Multimemory Sensor Quartz and the Speedmaster Professional.

The gold watch collection Joaillier was described in a press release as follows:

"The concept of the Omega gold collection is new with respect to the product as much as with respect to its marketing and presentation. The product can be distinguished by the longer lines, the softer forms, even thinner case and the accentuating of the white gold by the yellow gold. With respect to the presentation special attention was given to the boxes (red leather with harmonic shapes), the buckles for leather straps, with a gold screw replacing the pin, the alligator straps with hand made leather lining and waxed rims. For example gold watches with leather straps have hands which are brushed on one half and polished on the other half, which gives a very fine aesthetic. The numerals are in 18k gold. White gold looks more distinct when contrasted with some touches of yellow gold. The band is always color coordinated with the dial. Case and lugs are no longer cut in one piece, the Cornes Crosse or Corne Geneve, which are sol-

Omega De Ville Caliber 620 from 1967. This model began one of the most versatile collections in Omega's history.

97

dered to the case allowing lugs and case in corresponding decorations. More and more central attachments are favored which accentuate form and clear line of the case. One of those designs is the Retro-Replik of a model form 1936. The guilloched bezel is hand made by Omega crafts men. The bracelets have new braided bracelets, allowing for maximum flexibility and comfort, even for the smallest wrist. The polished bezel is revived, however with engraved sides. Two-tone models favor white gold over yellow gold combined with pink gold. The tri-color models have perfect harmony and integrated bracelets. Jewelry watches are usually in yellow gold, which contrasts better to diamonds (exclusively Top Wesselton VVS), it creates a warmer luster. In 1981 a new diamond setting is introduced (named "Pompadour"), where the diamonds are set in gold and internally held by two claws. The set pieces are either set in rows or around the bezel... As a speciality, the $20 coin watch is reintroduced in 1981."

When Omega became well known as a jewelry watchmaker, Italian counterfeiters decided that Omega was a good business opportunity. During the 1970s worthless imitations of valuable Omega models heavily set with diamonds were offered to tourists in parking lots. Those were well made imitations and many people were fooled and suffered large financial loss.

In 1982, a Constellation model was introduced where four claws seemed to hold the crystal. The Seamaster Titane, whose titanium case was highlighted with yellow gold accents, also received a new design. The case design with integrated bracelet continued to be one of the characteristics of the Seamaster collection.

A new look for Omega: The diver's watch Seamaster 300m from 1993. The pressure button is at 10 o'clock.

98

Pages 100 and 101: From the Louis Brandt collection with its distinguished case designs: the automatic model and the chronograph, both introduced in 1992.

In memory of the founder of the company, the Louis Brandt collection was launched in 1984. It offers only very expensive and complicated mechanical gold watches, which were hand made and decorated. All models have a sapphire crystal back, displaying the movement underneath. Some models are made in limited editions. In 1992 the Louis Brandt collection offered an automatic wristwatch with sweep seconds and date, a chronograph with 12 hour register, and a date and perpetual calendar watch with phases of the moon.

In 1991 the DeVille collection was expanded and now also contained models in 18k gold, to strengthen Omega's position in the market for prestigious watches.

The last five years have brought a number of anniversary watches: in 1989, a special edition Speedmaster Professional was created in memory of the 20th anniversary of the moon landing, and in 1992 the 50th anniversary of Caliber 27 CHRO was celebrated with an exclusive and expensive chronograph collection for a more affluent clientele.

Technical Features

Watch collectors are always looking for unusual models that are rare or unique and have technical peculiarities. A collector would be very excited to acquire an Omega with jump seconds or a self-winding watch with wind indication, which for Omega watches is a short hand at the center of the dial. The chronometer wristwatch 30 TS with dual subsidiary seconds will most likely remain a dream watch, because only one prototype exists. In addition to the subsidiary dials, the watch has an additional jumping sweep seconds. Highly collectible are the Omega watches with self-winding and regulator dials and watches with a push button at 8 o'clock, which allows one to stop the movement of the balance, but also stops the watch. Also mentioned at this time should be the Calendar watch, which has three apertures for giving the date, day of the week, and month digitally.

Very often, it is the watch collector who helps the manufacturer to shed light on its production history and solve the mysteries surrounding certain products.

<u>From 300 to 2,359,296 Herz</u>

Omega introduced electronic time keepers to the public in 1970. At the Basal fair, the Bieler company, up to this point considered the flag ship of Swiss watch industry, presented forty-eight pieces as forerunners to its Electroquartz f 8192 Hz series. The watches contained the Swiss Quartz Caliber Beta 21, which had been developed by the Centre Electronique Horloger in Neuenburg in cooperation with Omega. The movement received Omega No. 1300 (R 24.3 Q SC CAL STS PH 13P).

At this point in time, Omega also focused on the tuning fork Caliber developed by Ebauches SA. It was used in the chronometer wristwatch f 300 (a frequence of 300 Hz). This product was also later used in the chronograph wristwatch Speedsonic. In 1970 Omega also introduced the first high frequency wristwatch. The movement (Caliber 1500) made 2.4 million oscillations per second (!) and had been developed in cooperation with the Institute Batelle in Geneva. Several prototypes of the Megaquartz 2400 were shown at the Basel Fair. It would take another four years until the watch could be offered in small quantities.

Omega's own research was at a very hectic stage in 1970. Through its Holding company SSIH, Omega had taken over the fully electronic Pulsar watches from Hamilton, after the American watchmaker had run into financial difficulties. The result was the Time Computer for the wrist, introduced in 1972. It was equipped with a much improved diode indication, which, due to the high need of power, could not be read permanently. The movement bore number 1600 (35 Q LED CAL). The following is an article in the Austrian Watch and Jewelry Magazine of September about this novelty:

The Omega Chrono-Quartz Model with mixed display from 1976. It contains Caliber 1611.

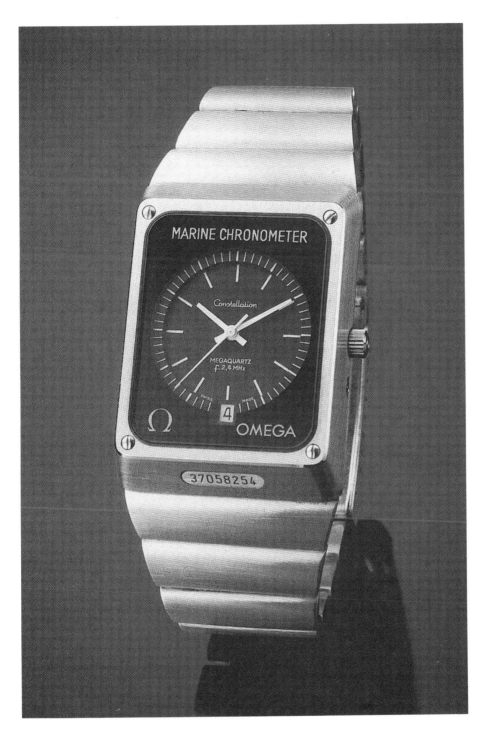

Omega's Marine Chronometer for the wrist from 1974. The Constellation
with Quartz Caliber F 2.4 MHz had sweep seconds and date. (In 1980 Omega
launched a Ships Chronometer with 4.19 MHz).

105

*The jewel watch Dinosaur (Caliber 1355) from
1980, an ultra thin quartz watch. The movement
is only 1.48 mm high. In 18k gold case with
alligator strap and gold buckle.*

The Magic Watch from 1981. The total height was 2.60 mm and was transparent at the center of the dial. It had two sapphire crystals and had a leather strap with gold buckle.

The Omega Time Computer is one of the first fully electronic watches with time indication by diodes. If you push the button for time, hour and minutes will be digitally displayed for 1.25 seconds. If you continue to press the button, hour and minute display will disappear and the seconds will be counted in front of your eyes. This futuristic watch makes use of space technology, which always requires a most perfect electric circle. They have neither crown, spring, hands or any other common watch part. All watch functions are fulfilled by one single little computer. Contrary to the first electronic watches, this watch has the advantage of being absolutely quiet.

The Omega Time Computer is fed by two batteries—which—for 25 readings per day will last for approximately 1 year. The quartz (32,768 vibrations per second) guarantees high precision (the maxi-

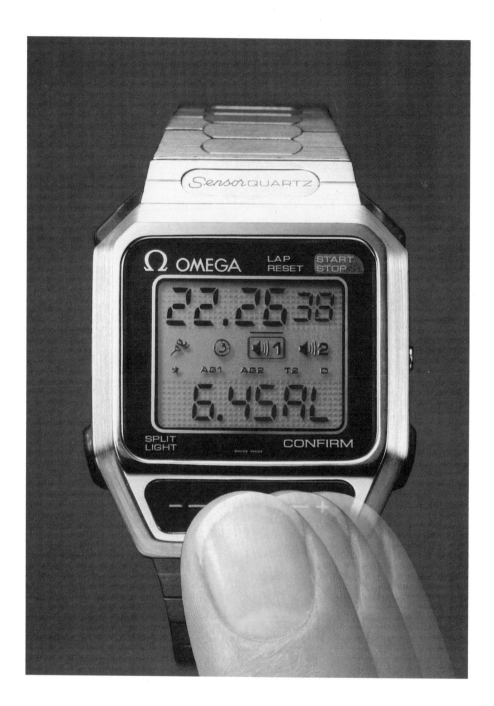

The sensor quartz from 1980 had several memory functions, which could be easily activated by pressing the appropriate key.

Multi-function watch Equinoxe from 1981, a reverse with dual dials—simple elegance with classic dial for the evening and contemporary digital technology for everyday.

mum deviation is 1 second per year). The very robust case is water-resistant, in order to protect the electronic parts from humidity and dust..."

The LED indication was soon replaced by LCD (Liquid crystal display).

While a standard mechanical watch has a frequency of 2.5 Hertz (= 2.5 oscillations per second = 18,000 half oscillations per hour) the Omega f 2,4 MHz had 2,359.296 oscillations per seconds (!), which are 2359 Kilohertz or 2.4 Megahertz. This gigantic frequency

109

Seamaster 120 m Quartz from 1981.

was divided by the electronic until the impulses were in one second intervals. The second hand makes one step per second, whereas with mechanical watches it moves 5 steps per seconds (= 5 half oscillations).

In 1993 Marco Richon wrote about the Super Constellation f 2,4 in a publication entitled *Omega—The History of a Great Name*: "the Megaquartz 2400 is by far the most precise wristwatch on earth...This thanks to a small disc form resonator, sealed into a capsule, which works with the unimaginable frequency of 2,359.296 oscillations per second. This high frequency has allowed the Caliber 1511 version to receive the classification of a Marine Chronometer—a unique distinction. After 63 days of the hardest testing, the watch was off by less than 2/1000 seconds per day! In addition the Megaquartz 2,400

Quartz models from the Seamaster collection from 1982. The scratch resistant surface of the case and band is made of argillaceous earth and titanium.

is the first watch with a TSA Display (Time Zone and Second Adjustment) which allows one to set the time at a second time zone, without interference with minute or seconds, and to synchronize the seconds. The first time in the history of watchmaking, the seconds had an important role to play."

The further modified model now was named No. 1510 (R 25,6 Q SCS CAL CORH CORS STS 13 P). The dimensions were 25.60 x 31 x 6 mm. The surface of 4 square mm carried 400 transistors. Construction following the building block system-quartz block with lentil shaped quartz disc, time indicating block. The marine chronometer had Omega number 1511 (OBS = Observatory certificate).

1974 brought the introduction of the first quartz model at Omega for analog time, Caliber 1310 (29Q SCS CALD CORR CORH CORS

111

STS 8P). The watch was named Megaquartz 32 KHz. Four of those Seamaster Megaquartz 32 KHz crossed the Atlantic on top of the mast and at the tip of the body of the two catamarans sailed by Ambrogio Fogar and the Sloop sailed by Paolo Mascheroni in 1976. They were the only two board instruments of the two Italians. The frequency of 32.768 Hz became the generally accepted frequency in the watch industry and the resonator was downsized to a miniature quartz stick. In 1976 another high technology for the wrist was created, the Chrono-Quartz (Caliber 1611) with dual indication. The time could be read analog, the chronograph was read digital, the stop movement measured 1/100 of a second. The watch had two batteries and 15 jewels. It was only a matter of time until the Speedmaster Professional was also made with quartz movement—it happened in 1977: Caliber 1620 (29Q LCD CALD CHROR C12 C60 c100) started to compete against mechanical watches.

In 1977 the small quartz caliber 1350 with the measurements 13 x 15.15 x 3.35 was introduced with a volume of 0.533 cm3. It opened new avenues for the jewelry watch designers. From 1978 on a baguette quartz movement with the measurements 9 x 21 x 3.35 was available. The volume of this Caliber (1352) was 0.613 cm3. In 1979 Omega launched its first LCD-lady's wristwatch with several memory options, the so-called Memomaster Quartz (Caliber 1637). The gentlemen's model with Caliber 1632 (29Q LCD CALD) was on the market since 1978.

In 1980 Omega added to its electronic wristwatch program with three technically sophisticated products. The ultra thin Dinosaur with Caliber 1355 and the Sensor Quartz, which was equipped with several micro processors, Caliber 1640.

The Dinosaur was, including the case, only 1.48 mm high—the height of several custom-made watches were as small as 1.35 mm. This was possible because the batteries and other parts of the watch were positioned outside of the dial. The hands seemed to float. Caliber 1355: R Q CORH CORM.

The Omega Sensor Quartz was a LCD wristwatch with memory function and nine programs and an easy to use sensor keyboard: a soft touch of the keys is sufficient to use various functions.

Seamaster Multi-Function, launched in 1988. Caliber 1665 is set in a titanium and gold case with integrated bracelet of a 1982 design.

113

At Omega, each new year saw a new breakthrough. The novelty in 1981 was the Magic Watch wristwatch with transparent dial and movement located outside of the dial (Caliber 1356/57) and the Equinoxe, a reversable watch with two dials. One side was the traditional dial, the other side had a multi-memory with display and LCD indication. The Caliber bore No. 1655 (R 20 Q CALD, CHROR, C 100 A 24, time sign, alarm, countdown).

The Seamaster collection also offered the 120m/Quartz, an elegant diver's watch with screwed crown, recessed crown, crystal, sweep seconds, and date. It contained Caliber 1337 (R 25,6 Q SCS CAL CORH CORM CORS IFP). This watch was tested for its every day use in the fall of 1981 by the French Jacques Mayol in the shore waters of the Elbe island. In a medical and technical experiment, the diver was pulled by a cast iron mass to a depth of 101 m without wearing an oxygen tank. The Seamaster was not only worn at this experiment, but it also played an important role. Mayol says: "When diving without oxygen tank, the objective time, which is measured with a time keeper does not correspond anymore to our subjectively felt inner time. The holding of breath seems to be accompanied by some kind of contraction of time, when the felt second is different from the clocked second. Therefore the watch has a vital task as an objective instrument, which counts the time and allows the diver to keep composure in the most difficult of situations."

The chronometer wristwatch Constellation was of course also offered with a quartz movement. The Quartz Constellation from 1982 had the famous crab setting with two claws at 3 and 9 o'clock, which also served as crown protection. The heart of the watch was Caliber 1422 (23,2 Q SCS CAL CORR STS IFB BULL 7P). On May 18th, 1983, Omega received the 100,000th chronometer certificate for a Quartz Chronometer.

The Seamaster Titane had Caliber 1422 (23,3 Q SCS CAL). The Caliber came from ETA, which after 1983 became more and more important as movement suppliers and eventually Omega stopped making movements.

The chronometer wristwatch Seamaster 1/100 from 1988, a quartz watch with traditional dial and time indication plus two LCD windows.

In the fall of 1986 Omega launched a new Seamaster Multi Function. Richon: "Eight dials and a magic crown. This is the most sophisticated and yet easy to use multi-function watch. The two mechanical hands give the local time, while at the same time an electronic screen liquid crystals can perform three traditional time functions. In addition, it gives four additional instrument functions: Count down, dual time zone, date, day of the week, stop watch, and alarm. Every function is activated by a single crown which can be turned forward and backwards to the appropriate function. Case and band are in titanium and 18k gold. Titanium is the preferred metal of the space program: half the weight of steel, yet the same strength and much more resistant to corrosion and scratches. Sapphire crystal. Water-resistant to a depth of 30 meters." All those functions were accomplished by Caliber 1665 (28 Q LCD AFA CALD CHRO 1/100 REV), which was listed by ETA as 988,431.

Shortly before the 1988 Olympic Games in Seoul, Korea, Omega introduced its Seamaster Chronograph 1/100. The watch contains 50,000 transistors and could never reach its technical limit. For example, the memory could time, during a car race, the time for each round—for up to 99 rounds, in 1/100 seconds and with partial memorization. The 1/100 seconds were made visible with a hand. The battery indicator warned of batteries starting to run low. This intelligent Caliber had No. 1670.

That same year, the Seamaster 200m was introduced, also offered in Quartz movement, with Caliber 1438 (23,3 Q SCS IFB ASS).

In 1986 Omega launched two new product lines, the "Symbol" and the "Art."
The distinguishing characteristic of the Symbol was its colorful dial and the
symbolic images from Egypt, India, and China (e.g., Yin and Yang). The Art was
a Quartz wristwatch with simple aesthetics and a case with new alloys. The
back case bore an exclusive Concrete-Art medallion with the artists signature.
The illustrated model from 1987 came in a limited edition of 999 pieces.

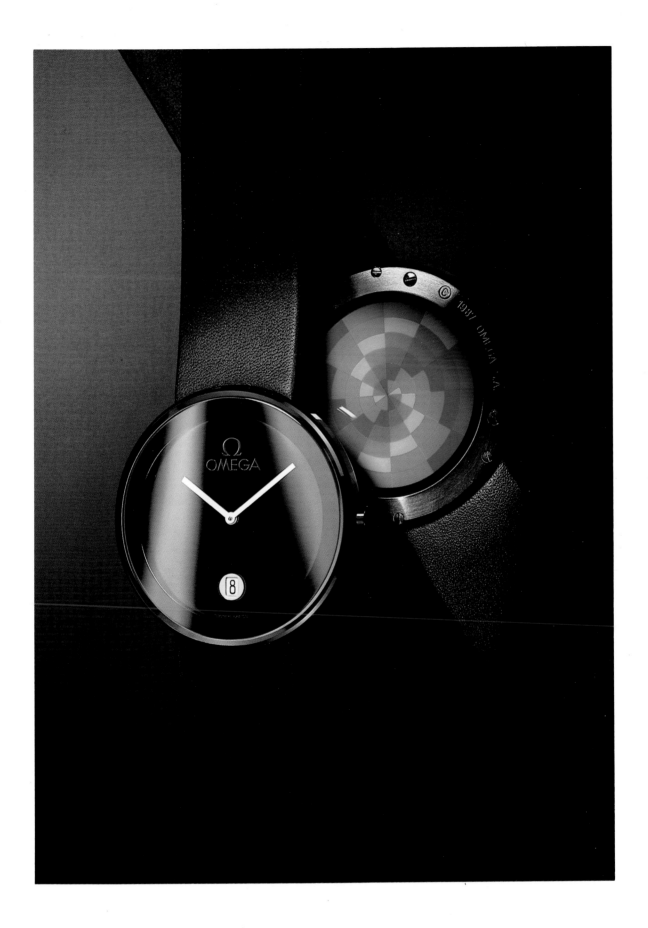

The ultra thin automatic movement from 1942, Caliber 30,10
with swinging mass and exposed bumper springs (hammer
automatic), After 1949 it bore Caliber No. 330.

List of Calibers and Movement Numbers

Because of their high movement quality, the diligent regulation and its solid cases, which together kept the balance from losing all but a very small oscillation radius each year, Omega watches became famous within a short period of time. Part of this reputation was based on the fact that Omega used exclusively first quality movement parts. During the economic crisis following the First World War, parts of lesser quality had to be used, but by 1923 Omega had returned to first quality parts. A Breguet hairspring, for example, was part of the standard edition of an Omega Caliber.

By the end of the 1930s Omega movements received a pink gilding as surface protection, which was continued and became an optical trademark for Omega.

In 1928 the Bieler watchmaker Henri Gerber became Omega's new technical director, and he would guide the company's development for the next thirty-seven years. He was the mastermind of the technical evolution of these decades. During his leadership, the first water-resistant wristwatches, self-winding watches, and chronometer wristwatches were developed. In 1939 he introduced the famous manual Caliber 30 mm. Gerber achieved exceptional results through improvements of the movements and the automation of production.

One of the most important improvements for wristwatches was the addition of a sufficient protection against balance staff breakage caused by shock or a drop from height. From 1940 on, Omega built shock resistant watches and in 1943 decided on the Incabloc resistance, which had been fully developed in 1938. The balance jewels are housed in a conical setting, and if the intensity of the shock or push was above norm, the jewels would give way to the pressure until the robust part of the balance wheel was touching the plane, thus absorbing the shock. After that the pressure from the retaining spring brought the system back to its original position. The Incabloc spring was in the shape of a lyre. The automatic recentering took place with such precision that the operation of the movement was not interrupted. The oil reserve was kept in a dust-proof chamber, which was the basis for assuring the lubrication.

The next important improvement of the mechanical wristwatch was the Nivarox spring in the 1930s. Its development eliminated four problems which watchmakers had to deal with: magnetism, temperature, elasticity, and oxidation. Reinhard Straumann had found an alloy which could compensate for all those things. The alloy consisted of iron, nickel, beryllium, titanium, and silicious. Omega instantly retired the steel spring and put the new spring into its movement, usually as overcoil springs. The Breguet spring was redundant only after 1960.

The Nivarox-spring also brought a new balance, replacing the bi-metallic compensation balance. Since the alloy's main component was beryllium (*glucydur* in French), it was called glucydur balance. Even though the new mono-metallic balance did not need any regulation screws, because of tradition they were still put on until the middle of the 1960s.

Omega certainly made good use of these technological advances and used a four arm glucydur balance with a sand-blasted gilded surface in its movement. Its price was about four times higher than a traditional balance, but the high quality standards made it indispensable. The traditional balance could be made in four steps whereas the glucydur balance required twenty-one. It had an unbelievable strength, allowing it to keep its original shape and be completely antimagnetic.

Another important issue for watchmakers was the design of the sweep seconds. Omega favored the so-called indirect version, the reason being the advantage of having the seconds gear running outside of the flow of energy from the spring barrel to the balance. The third wheel is connected to both the regular seconds gear and the center seconds gear. The latter is sitting in a cartouche which is pressed into the cam of the minute wheel. (This is, of course, a difficult to access location and is very delicate, and should not be exposed to large pressure). It should be noted that the caliber with indirect center seconds can also be used without significant change for watches with subsidiary seconds. Furthermore, this construction allows the balance wheel to sit close to the rim of the plate and therefore is easy to access and observe. It should also be noted that if the seconds wheel is within the power stream, it is set in jewels, while indirect seconds are not.

Every technically significant development since the 1930s dealt with similar problems: water-resistant cases, antimagnetism or shock resistant balances. Another one of these reoccurring problem areas was the self-winding mechanism. At the beginning of the 1940s, this issue could not be postponed any longer and had to be dealt with. In its earlier automatic calibers, Omega used the so-called hammer automatic, and it was not until 1955 that a rotor winding entered the Omega Caliber 455 (lady's caliber) and 470 (gentlemen's) production. In developing these new generations of calibers, Omega went its own way, which is reflected in the patents received. For example, Omega Caliber 455 had a winding mechanism with rocking arm and cam gear. Caliber 470 automatic winding had two wheels fitted on studs on the rocking arm. Omega also received several patents for its solution to free running conversion and for radial clicks. A technical novelty in that area was Caliber 550, which used the studs of two satellite wheels to fulfill the click function.

Each Omega watch movement was inscribed with a letter and number combination. The first numbers gave the movement diameter in millimeter, and the letters marked the particular characteristics for the watch. These caliber letters can be found somewhere around the balance.

In 1949 Omega decided to change this system in order to simplify and clarify it, and a three digit number for Caliber and the Omega letter were engraved on a bridge or plate. For its internal records, however, Omega continued the old system. For example: in 1952 the Caliber No. 420 was given to a movement 26,5 SC (movement diameter 26.5mm, sweep seconds).

This movement marking simplified the ordering of spare parts for a watchmaker, because it allowed for easy Caliber identification. For the collector the identification is a helpful aide for identifying a newly acquired watch. The following are the most common abbreviations (including quartz caliber):

121

A24	= 24 hour dial
AL	= Alarm
AM	= antimagnetic
AM/PM	= the time from 0 to 12 hours/12 to 24 hours (English indication)
AU	= Wristwatch
B	= quality rating for Omega Caliber: fifteen jewel lever movement
Bs	= Subrating for quality: higher than B
BULL	= rating certificate
C	= chronograph
C12	= 12 hour register
C24	= 24 hour register
C60	= 60 minute register
C100	= 1/100 seconds stop
CAL	= Calendar
CALD	= Calendar with date
CALP	= Perpetual calendar
CH	= Chronostop
CHRO	= Chronograph
CHROR	= Split second chronograph
CORH	= Hour Set
CORJ	= Day Set
CORM	= Minute Set
CORR	= Date Set
CORS	= Seconds Set
CR	= Double hand register
DL	= Date and Phases of the moon
ECL	= Illumination
EP	= ultra thin
ETANCHE	= water-resistant
F	= Form movement
FU	= Time zones
GMT	= Greenwich Mean Time
H	= Movement Height
IFP	= Battery charge indication
JUB	= Jubilee Caliber
L	= Openface pocket watch
LCD	= Liquid Crystal Display
LED	= Light Emitting Diode

LINIE ('''')	= old measure for length, measure for Caliber; I Linie (ligne) = 2.256 mm
MO	= Month
P	= Plate
P	= jewels
PC	= shock resistant
PS	= subsidiary seconds
Q	= Quartz
R	= Form movement
R	= Double hand
RA	= automatic winding
REM	= manual winding
REV	= alarm
RG	= index regulation
RH	= hour signal
RS	= tuning fork
S	= hunter cased watch
S	= Seconds
SAV	= Hunter cased watch
SC	= indirect seconds
SCD	= direct seconds
SCS	= jumping center seconds
SIZE	= American Calibers
SON	= Repeating
STS	= stop seconds
T	= tonneau movement form
T I	= First modification of movement
T2	= Second modification of movement
T3	= Third modification of movement
TM	= Parking time alarm

Originally the movements were signed with Omega in letters, but once in a while instead of Omega the Greek letter Ω was signed. Later the signature OMEGA WATCH CO SWISS was used.

Along with the Caliber Number, each movement also had a production number. A look in the movement number file of Omega might be quite confusing, because Omega was not very diligent in keeping numerical order. Different series of numbers were used simultaneously for many years, which means that a watch with a higher movement number might not necessarily be younger, or one with a lower num-

123

ber older. Movement numbers starting with 5 or 9 million can not be timed at all.

Despite these inconsistencies, I do not want to disregard the movement number list for the seventy-five year period from 1894 to 1969. It should be noted, however, that it is not more than a vague reference. For example, in 1894 the watch with movement number 1,000,000 was delivered, much ahead of the actual production. Movement with special numbers, such as full millions, were reserved for special clients or went to members of the Brandt family. Following the numbers, the date up to which those movement numbers were still used in small series is indicated in parentheses. In combination with the Caliber number, at least an approximate dating of a particular watch is possible with these numbers.

Ω Movement Numbers	Produced
- 1,999.999	to 1902 (1916)
- 2,999.999	to 1908 (1919)
- 3,999.999	to 1912 (1919)
- 4,999.999	to 1916 (1927)
- 5,999.999	to 1923 (1927)
- 6,999.999	to 1929 (1935)
- 7,999.999	to 1935 (1941)
- 8,999.999	to 1939 (1944)
- 9,999.999	to 1944 (1950)
-10,999.999	to 1947 (1951)
-11,999.999	to 1950 (1953)
-12,999.999	to 1952 (1955)
-13,999.999	to 1954 (1957)
-14,999.999	to 1956 (1958)
-15,999.999	to 1958 (1962)
-16,999.999	to 1961
-17,999.999	1961 (1963)
-18,999.999	1963 (1964)
-19,999.999	1963
-20,999.999	1964 (1967)
-21,999.999	1965 (1966)

-22,999.999	1966 (1969)
-23,999.999	1968
-24,999.999	1967 (1969)
-25,999.999	1969
-26,999.999	1968 (1970)
-27,999.999	1969 (1970)
-28,999.999	1969 (1970)

During the 1980s Omega stopped producing its own movements, which were replaced by ETA Calibers.

The following lists important Calibers in chronological order:

Cal. 15 lig.
13 lig.
12 lig.
11 lig.
10 lig.

These five movements (diameters 33.84 mm, 29.33 mm, 27.07 mm, 24.81 mm and 22.56 mm) came on the market in 1898 and were smaller editions of the 19 ligne Omega Caliber from 1894. The circular lever movements with 3/4 plates were also used in the early wristwatches at the turn of the century. The 12 ligne movement was eliminated from production, because it was not successful for Omega.

C. 57. Mouvement Lépine 10''',
qualité B.

C. 331. Mouvement Lépine 11''',
qualité Bˢ.

C. 78. Mouvement Lépine 12''',
qualité Bˢ.

C. 311. Mouvement Lépine 13''',
qualité B.

C. 79. Mouvement Lépine 15'''.
qualité B.

125

cal. 17 lig.

Circular Lever movement with 3/4 plate, diameter 38.35 mm, cut bimetallic compensation balance, subsidiary seconds, signed on balance plate, gilded movement. Similar to the other movements of this first Caliber generation, this Caliber had the modern Omega winding (no more setting pin at winding and setting crown). These large pocket watch movements from 1898 were sometimes used for large wristwatches and observation watches for the wrist.

C. 80.
Mouvement Lépine 17''', qualité B.

cal. 9 lig.

Circular lever movement, diameter 20.30 mm, height 3.70 mm, 18 jewels (!), "qualite speciale." Came on the market in 1914. It was mainly developed for good lady's watches. It is mainly found until 1925 in various lady's and gentlemen's watches and in large rectangular wristwatches. In 1923 it was replaced by Cal. 19,4.

Cal. 9'''
18 pierres

Models from 1925 with a Caliber 9 lig. movement.

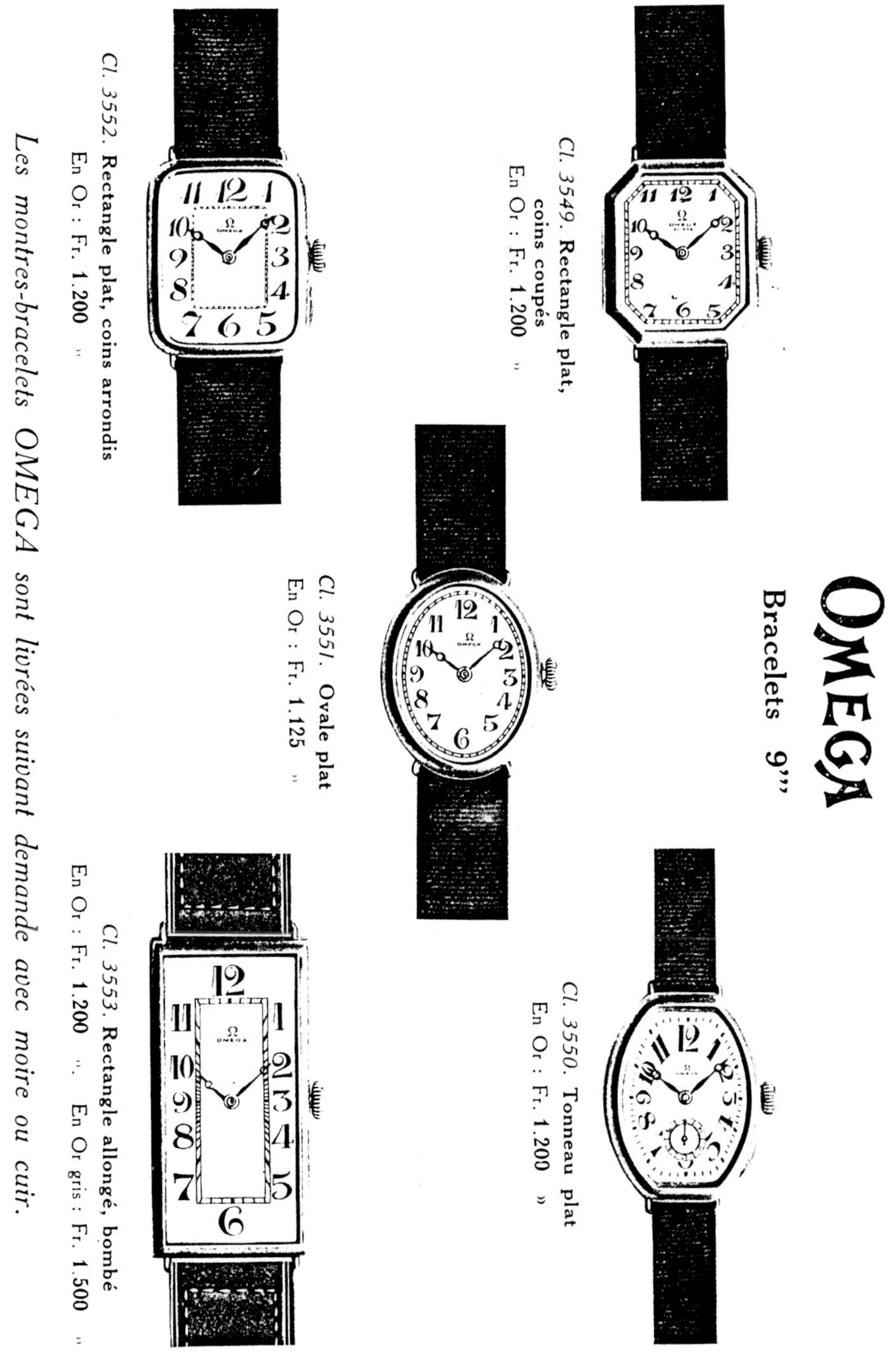

OMEGA

Bracelets 9"'

Cl. 3549. Rectangle plat,
coins coupés
En Or : Fr. 1.200

Cl. 3552. Rectangle plat,
coins arrondis
En Or : Fr. 1.200   "

Cl. 3551. Ovale plat
En Or : Fr. 1.125   "

Cl. 3550. Tonneau plat
En Or : Fr. 1.200   "

Cl. 3553. Rectangle allongé, bombé
En Or : Fr. 1.200   ".   En Or gris : Fr. 1.500   "

Les montres-bracelets OMEGA sont livrées suivant demande avec moire ou cuir.

127

Cal. 3/0 SIZE   Circular lever movement, diameter 27.94 mm.
                  Since 1917.

Cal. 10 1/2 lig. Circular lever movement, diameter 23.69 mm.
                  Since 1917.

Cal. 23,7   Circular lever movement, silvered 3/4 plate, di-
                  ameter 23.69 mm, height 3.80 mm, 15 jewels,
                  subsidiary seconds. Since 1918. The movement
                  was modified several times (T1, T2, T3).

Cal. 23,7   Cal. 23,7   Cal. 23,7   Cal. 23,7
15 pierres   15 pierres   T. 1 ~ T. 2 ~ ou T. 3   T. 1 ~ T. 2 ~ ou T. 3
                                      15 pierres   15 pierres

Cal. 35,5   Thin lever movement, diameter 35.5 mm, height
                  4.60 mm, 15 jewels, Breguet spring. The Caliber
                  was developed in 1918 for thin pocket watches
                  and occasionally was used for early aviator's wrist-
                  watches.

Pages 129 to 133: Gentlemen's wristwatches with circular or form movements,
Caliber 23,7 from circa 1923.
Pages 134 to 136: Wristwatches with Caliber 23,7 from the 1925 collection.
Pages 137 to 138: Wristwatches with Caliber 23,7 and 19,4 from 1931.

Genève 23,7/12'''
133. MA 712 Metal
134. AR 712 Plata 0,925
135. DB 712 Chapeado de oro 10 años
136. OT 712 Oro 18 Kt.

Empire 23,7/12'''
137. AR 780 Plata 0,925, joncs blancs
138. OT 780 Oro 18 Kt.

Empire, joncs ciselés 23,7/12'''
139. AR 782 Plata 0,925
140. OT 782 Oro 18 Kt.

Monnaie 23,7/12'''
141. AR 784 Plata 0,925
142. OT 784 Oro 18 Kt.

Carrée cintrée 23,7/12'''

143. MA 800 Metal
144. AR 800 Plata 0,925
145. DB 800 Chapeado de oro 10 años
146. OT 800 Oro 18 Kt.

Biseau, carrure ciselée 23,7/12'''

147. AR 813 Plata 0,925
148. DB 813 Chapeado de oro 10 años
149. OT 813 Oro 18 Kt.

Double biseau, carrure plate à blocs 23,7/12'''

150. AR 814 Plata 0,925
151. OT 814 Oro 18 Kt.

Paris, lunette jonc, 23,7/12'''

152. AR 817 Plata 0,925
153. OT 817 Oro 18 Kt.

Rectangle-losange à biseau à bloc 23,7 mm.
154. AR 721 Plata 0,925
155. OT 721 Oro 18 Kt.

Tonneau allongé à biseau à bloc 23,7 mm.

156. AR 759 Plata 0,925
157. DB 759 Chapeado de oro 10 años
158. OT 759 Oro 18 Kt.

Rectangle à biseau, coins vifs à bloc 23,7 mm.

159. AR 720 Plata 0,925
160. DB 720 Chapeado de oro 10 años
161. OT 720 Oro 18 Kt.

Rectangle bombé 23,7 mm.
162. AR 727 Plata 0,925
163. DB 727 Chapeado de oro 10 años
164. OT 727 Oro 18 Kt.

Tonneau losange à biseau, à blocs 23,7 mm.

165. AR 775 Plata 0,925
166. DB 775 Chapeado de oro 10 años
167. OT 775 Oro 18 Kt.

Tonneau tronqué à biseau, à blocs grecs fantaisie 23,7 mm.

168. OT 779 Oro 18 Kt.

Carrée à biseau, coins vifs à blocs 23,7 mm.

169. AR 797 Plata 0,925
170. DB 797 Chapeado de oro 10 años
171. OT 797 Oro 18 Kt.

Carrée à biseau à blocs grecs, 23,7 mm.
172. OT 799 Oro 18 Kt.

Carrée à biseau, coins coupés à blocs pointus, 23,7 mm.
173. OT 863 Oro 18 Kt.

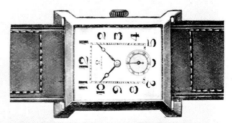

Rectangle à biseau, à blocs grecs 23,7 mm.
174. OT 870 Oro 18 Kt.

Octogone rectangle à biseau à blocs 23,7 mm.
175. AR 833 Plata 0,925
176. DB 833 Chapeado de oro 10 años
177. OT 833 Oro 18 Kt.

OMEGA
Bracelets 23,7 m/m

Cl. 3569. Carrée à biseau, coins coupés
En Argent . . . . Fr. 550 »
En Or . . . . . Fr. 1.300 »

Cl. 3570. Tonneau tronqué à biseau
En Argent . . . Fr. 500 »
En Or . . . . . Fr. 1.000 »

Cl. 3571. Rectangle bombé
En Argent . . . Fr. 500 »
En Plaqué Or. . Fr. 525 »
En Or . . . . . Fr. 1.100 »

Cl. 3572. Carrée, coins coupés
En Argent : Fr. 480 »   En Or : Fr. 950 »

Cl. 3573. Tonneau tronqué
En Argent : Fr. 480 »   En Or : Fr. 950 »

Tous les bracelets peuvent être livrés avec cadran lumineux au radium, augmentation Fr. 25 »

OMEGA

Bracelets 23,7 m/m

Cl. 3574. Genève, serrée
Argent : Fr. 375 "
Plaqué Or : Fr. 400 " Or : Fr. 750 "

Cl. 3575. Biseau, Carrure plate
En Or : Fr. 800 "

Cl. 3576. Rectangle, plat, cintré
Argent . . . . : Fr. 480 "
Or . . . . : Fr. 1.000 "

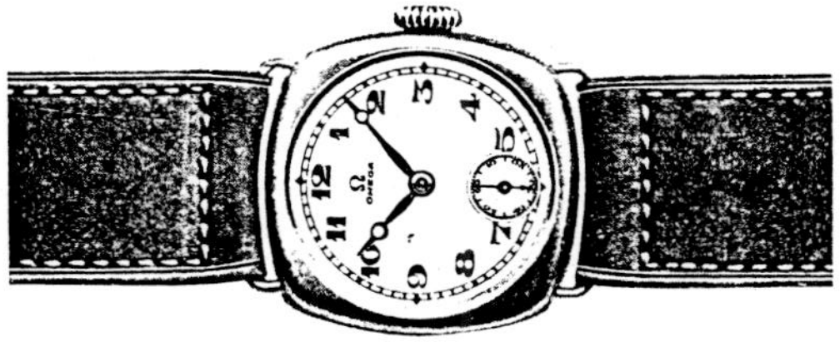

Cl. 3577. Carrée cintrée
Argent : Fr. 420 " Plaqué Or : Fr. 445 "
Or : Fr. 900 "

Cl. 3578. Lunette biseau
En Plaqué Or : Fr. 400 "

La montre-bracelet ne remplace pas la montre de poche, elle la complète.

135

OMEGA

Bracelets 23,7 m/m

Cl. 3579. Monnaie
En Plaqué Or . . . Fr. 400 »
En Or . . . . Fr. 750 »

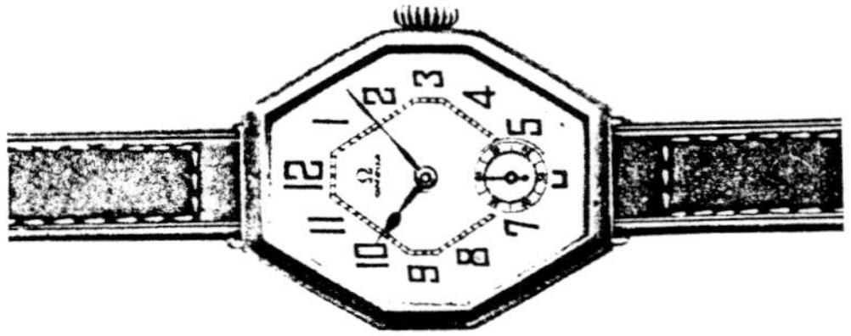

Cl. 3582. Octogone-tonneau
En Or : Fr. 1.000 »

Cl. 3581.
Carrée, lunette à biseau
En Or : Fr. 1.000 »

Cl. 3580. Octogone
En Or . . . . Fr. 850 »

Cl. 3583. Curviligne
En Or : Fr. 1.100 »

Quelle que soit sa forme, la montre-bracelet OMEGA est aussi robuste que la montre de poche.

OMEGA

Gents' Wristlet Watches
Model 237   15 Jewels

Cushion, solid loops
shaped glass :

Silver	£ 5 . 15 . 0
9 ct. solid Gold	£ 8 . 5 . 0

Rectangular :

Silver	£ 6 . 0 . 0

Tonneau :

Silver	£ 6 . 12 . 6
9 ct. solid Gold	£ 10 . 10 . 0
18 ct. solid Gold	£ 12 . 10 . 0

Screw Bezel, Swing Ring :
9 ct. solid Gold   £ 7 . 15 . 0

Round Tonneau, Enamel Figures :

9 ct. solid Gold	£ 8 . 0 . 0
Silver	£ 5 . 12 . 6

Screw Bezel
Swing Ring, open

Cushion, Enamel Zone :

Silver	£ 6 . 0 . 0
9 ct. solid Gold	£ 8 . 10 . 0

10   Ω

137

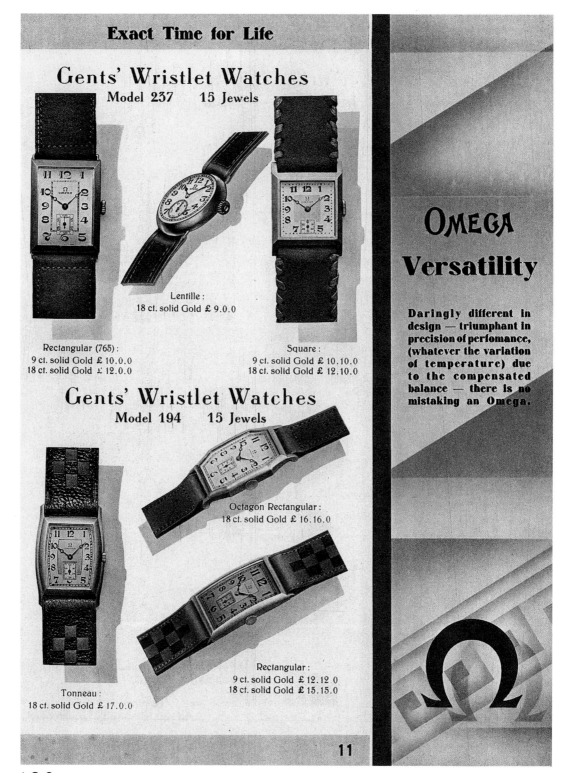

Exact Time for Life

Gents' Wristlet Watches
Model 237   15 Jewels

Lentille :
18 ct. solid Gold £ 9.0.0

Rectangular (765) :
9 ct. solid Gold £ 10.0.0
18 ct. solid Gold £ 12.0.0

Square :
9 ct. solid Gold £ 10.10.0
18 ct. solid Gold £ 12.10.0

Gents' Wristlet Watches
Model 194   15 Jewels

Octagon Rectangular :
18 ct. solid Gold £ 16.16.0

Rectangular :
9 ct. solid Gold £ 12.12.0
18 ct. solid Gold £ 15.15.0

Tonneau :
18 ct. solid Gold £ 17.0.0

OMEGA Versatility

Daringly different in design — triumphant in precision of perfomance, (whatever the variation of temperature) due to the compensated balance — there is no mistaking an Omega.

11

138

Cal. 12,3

Oval lever movement, diameter from 3 to 9 o'clock is 12.30 mm, ultra thin (2.70 mm), silvered movement plate, 17 jewels, Breguet spring, first quality finishing (= "Geneve"), covered winding wheels, adjusted to two positions. "With the Caliber 12,3 F (5 lign,) we are producing watches, whose elegance and quality is hard to surpass. These small and thin calibers can be found in many different case styles, and are offered at all price levels." It was put on the market in 1922.

Cal. 12,3   Cal. 12,3

Cal. 19,4

Circular lever movement with silvered 3/4 plate, diameter 19.40 mm, height 3.15 mm, Breguet spring, 15 jewels, subsidiary seconds; fine finish, adjusted to two positions. Was modified several times (1930 T1, 1935 T2). With... "we are launching a new caliber 19,4 (8 3/4'")", a true wonder of precision". It was offered beginning in 1923.

Calibre 19,4 T 2

Page 140: Elegant wristwatch with Caliber 12,3 (circa 1923).
Pages 141 to 145: Gentlemen's and lady's watches with Caliber 19,4 from the 1923 collection.
Page 146: Lady's watches with Caliber 19,4 from 1931.
Pages 147 to 148: Gentlemen's watches with Caliber 26,5 from 1931.
Page 149: Lady's wristwatch with Calibers 14,8, 12,3 and 2,5 from 1931.

Ovale 12,3 F. mm.
198. OT 822 Oro 18 Kt.
199. OG 822 Oro blanco. 18 Kt.
200. PA 822 Platina

Rectangle coins coupés 12,3 F. mm.
201. OT 725 Oro 18 Kt.
202. OG 725 Oro blanco 18 Kt.
203. PA 725 Platina

Tonneau 12,3 F. mm.
204. OT 754 Oro 18 Kt.
205. OG 754 Oro blanco 18 Kt.
206. PA 754 Platina

Rectangle coins vifs 12,3 F. mm.
207. OT 728 Oro 18 Kt.
208. OG 728 Oro blanco 18 Kt.
209. PA 728 Platina

Rectangle 12,3 F. mm.
210. OT 726 Oro 18 Kt., con esmalte

Tonneau 12,3 F. mm.
211. PA 754 Platina con brillantes

Rectangle coins coupés 12,3 F. mm.
212. PA 725 Platina con brillantes

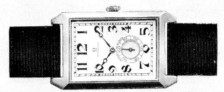

Rectangle à biseau, coins vifs 19,4 mm
178. OT 720 Oro 18 Kt.

Tonneau tronqué, à biseau, évasée à blocs 19,4 mm.
179. OT 761 Oro 18 Kt.

Carrée, lunette à biseau, coins vifs à blocs 19,4 mm.
180. OT 797 Oro 18 Kt.

Rectangle à biseau, évasée à boules 19,4 mm.
181. OT 871 Oro 18 Kt.

Tonneau festonné 19,4 mm.
186. OT 776 Oro 18 Kt.

Empire, carrure cannelée 19,4 mm.
187. OT 786 Oro 18 Kt.

Carrée cintrée, glace de forme 19,4 mm.
188. OT 801 Oro 18 Kt.

Croix de Malte 19,4 mm.
189. OT 795 Oro 18 Kt.

Carrure plate, lunette à biseau 19,4 mm.
182. OT 716 Oro 18 Kt.

Grand guichet, carrure ronde 19,4 mm.
183. OT 747 Oro 18 Kt.

Tonneau tronqué 19,4 mm.
184, OT 756 Oro 18 Kt.

Monnaie 19,4 mm
185. OT 783 Oro 18 Kt.

Carrée cintrée 19,4 mm.
190. OT 800 Oro 18 Kt.

Carrée, coins coupés 19,4 mm.
191. OT 803 Oro 18 Kt.

Paris, lunette jonc 19,4 mm.
192. OT 817 Oro 18 Kt.

Octogone 19,4 mm.
193. OT 830 Oro 18 Kt.

Ronde festons 19,4 mm.
194. OT 827 Oro 18 Kt.

Carrée cintrée à biseau 19,4 mm.
195. OT 860 Oro 18 Kt.

Octogone 19,4 mm.
196. OT 831 Oro 18 Kt.

Carrée cintrée, profil empire 19,4 mm.
197. OT 861 Oro 18 Kt.

OMEGA

Ladies' Wristlet Watches
Model 194　　15 Jewels

Paris :
9 ct. solid Gold　£ 8 . 0 . 0
18 ct. solid Gold　£ 9 . 9 . 0

Cushion Mirage :
9 ct. solid Gold　£ 8 . 0 . 0

Cut Corner Mirage :
9 ct. solid Gold　£ 8 . 0 . 0

Octagon Mirage :
9 ct. solid Gold　£ 8 . 0 . 0

Centric :
9 ct. solid Gold　£ 8 . 0 . 0

Cushion :
9 ct. solid Gold　£ 9 . 9 . 0
18 ct. solid Gold　£ 11 . 0 . 0

Cushion, Shaped Glass :
9 ct. solid Gold　£ 10 . 0 . 0

Tonneau :
9 ct. solid Gold　£ 10 . 0 . 0
18 ct. solid Gold　£ 11 . 10 . 0
(with Moiré Band only)

Octagon :
9 ct. solid Gold　£ 10 . 0 . 0
18 ct. solid Gold　£ 11 . 10 . 0
(with Moiré Band only)

14　　Ω

OMEGA
Gents' Wristlet Watches
Model 265   15 Jewels

Cushion, solid loops, round glass:
Silver . . . . . . £ 5 . 0 . 0
9 ct. solid Gold . . £ 7 . 12 . 6
18 ct. solid Gold . . £ 11 . 10 . 0

Paris:
Silver . . . . . . £ 4 . 12 . 6
20 Year Gold filled £ 4 . 12 . 6

Crystal Dome:
Silver . . . . . . £ 4 . 17 . 6
9 ct. solid Gold . . £ 7 . 5 . 0

Screw Bezel,
Swing Ring:
Silver . . . . £ 4.17.6
9 ct. solid Gold £ 7 . 5 . 0

Cushion, solid loops,
Shaped glass:
Silver . . . . . . £ 5 . 7 . 6
9 ct. solid Gold . . £ 8 . 0 . 0

Screw Bezel,
Swing Ring, open.

Cushion:
Chromium . . . . £ 4 . 0 . 0
Silver . . . . . . £ 4 . 17 . 6
20 Year Gold filled £ 4 . 17 . 6
9 ct. solid Gold . . £ 7 . 5 . 0
18 ct. solid Gold . . £ 10 . 10 . 0

8   Ω

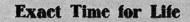

Exact Time for Life

Gents' Wristlet Watches
Model 265   15 Jewels

Lentille :
9 ct. solid Gold . . £ 5 . 17 . 6

Biseau :
20 Year Gold filled   £ 4 . 12 . 6

Gents' Wristlet Watches
Model 237   15 Jewels

Round Tonneau :
Silver . . . . . . £ 5 . 7 . 6
9 ct. solid Gold . . £ 7 . 15 . 0

Cushion :
Silver . . . . . . £ 5 . 7 . 6
9 ct. solid Gold   £ 7 . 15 . 0
18 ct. solid Gold . . £ 10 . 0 . 0

Unless otherwise ordered, all our 9 ct. Gold Gents' Wrist Watches are supplied with best quality Rolled Gold buckles, and 18 ct. Gold Gents' Wrist Watches with 9 ct. Gold buckles.

9

OMEGA Reliability

Abreast of the Fashion-correct with the time.

Reliable in every detail, Omega Watches bring satisfaction to a discriminating clientele.

The Case and Movement, which represent the finest workmanship, are sturdy and flawless.

An Omega will survive many a jar which would be fatal to another Watch.

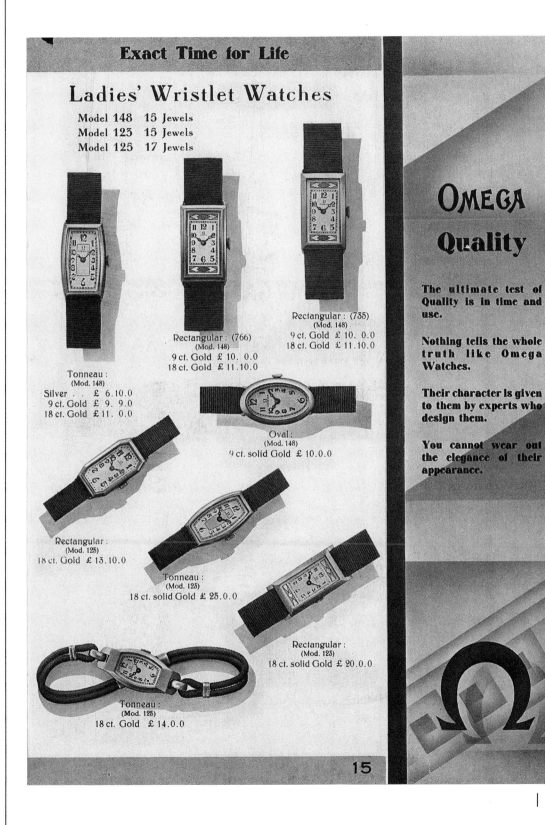

Exact Time for Life

Ladies' Wristlet Watches

Model 148   15 Jewels
Model 123   15 Jewels
Model 125   17 Jewels

Rectangular : (735)
(Mod. 148)
9 ct. Gold £ 10. 0.0
18 ct. Gold £ 11.10.0

Rectangular : (766)
(Mod. 148)
9 ct. Gold £ 10. 0.0
18 ct. Gold £ 11.10.0

Tonneau :
(Mod. 148)
Silver . . £ 6.10.0
9 ct. Gold £ 9. 9.0
18 ct. Gold £ 11. 0.0

Oval :
(Mod. 148)
9 ct. solid Gold £ 10.0.0

Rectangular :
(Mod. 125)
18 ct. Gold £ 13.10.0

Tonneau :
(Mod. 125)
18 ct. solid Gold £ 25.0.0

Rectangular :
(Mod. 123)
18 ct. solid Gold £ 20.0.0

Tonneau :
(Mod. 125)
18 ct. Gold £ 14.0.0

15

OMEGA Quality

The ultimate test of Quality is in time and use.

Nothing tells the whole truth like Omega Watches.

Their character is given to them by experts who design them.

You cannot wear out the elegance of their appearance.

Cal. 26,5

Circular lever movement, diameter 26.50 mm, 15 jewels. This movement replaced the 12 ligne Omega Caliber and was produced from 1926 on. It received several modifications, and after 1949 was renamed Caliber No. 100. A Caliber with incredible success. Some editions of this caliber were made with screwed bearings. The Caliber 26,5 T2 PC 15 was the first Omega movement with shock resistance.

Cal. 26,5
15 pierres

Cal. 26,5
15 pierres

100
15 pierres Jewels

Cal. 12,5 F T1

Oval lever movement, diameter from 3 to 9 o'clock is 12.50 mm. 1927.

Cal. 12,5

Cal. 12,5

Cal. 14,8

Cal. 14,8

Cal. 14,8

The larger version of Caliber 12,5. 1927.

Cal 39 CHRO
(130)

Since 1928 a new chronograph caliber was on the market. It was used for cars, airplanes, pocket, and wristwatches. Diameter 39 mm, height 7.15 mm, 17 jewels, chronograph mechanism with switching wheel, 30 minute register, one button for start, stop, and return, gilded movement, and adjusted to two positions. The movement was manufactured in Biel, the dial was supplied by LeCoultre in Le Sentier. Later the chronograph mechanism was also purchased from there; from 1932 on it came

again from Omega. With the introduction of the
new Caliber numbering in 1949, this caliber became No. 130.

130
17 pierres *Jewels*

Cal. 20

Tonneau lever movement, width 20 mm, winding
wheels under the bridge. Since 1928. A very successful movement.

„New Calibre" **OMEGA** This movement, calibre 20 m/m
F. is of quite recent creation
and is especially suitable for rectangular and barrel shaped watches.
This movement is already highly appreciated by our world-wide
customers, for its wonderful timekeeping qualities, sturdiness, elegance, style and moderate price. The following wristlet watches of
this catalogue have the illustrated movement: Nos 229, 230, 233,
236, 237, 238, 239, 240

229. Stainless-steel wristlet,
    rectangular   Rs 90

Remarks on stainless-steel

This is a steel the alloy of which contains a proportion of chromium: the
chromium is therefore not a coating
which is spread over the base steel,
but it is part of the steel alloy itself;
although we have just started to use
stainless-steel for watch cases, the tests
made have proved that the rustless
steel we are using is truly « rustless »
as its name implies

Shaped to the wrist

Above and following page: Gentlemen's wristwatch with Caliber 20, early 1930s.   151

OMEGA

EXACT TIME FOR LIFE

Shaped to the wrist

230. Silver chromium-plated wristlet,
rectangular   Rs 95

231. Silver wristlet, rectangular,
extra flat   Rs 110

Shaped to the wrist

232. Nickel chromium-plated
wristlet, barrel   Rs 60

Chromium-plating. We all know that silver and nickel tarnish and need constant cleaning; further, in hot weather and tropical countries, silver becomes rapidly black — not so when it is chromium plated — it always remains bright.

Chromium-plating on nickel is quite a recent process and more recent still on silver. For chromium plating nickel, the Omega case factory has installed special copper and nickel baths for depositing on watch cases a preliminary thick nickel coating. It is on this coating that the chromium plating is then applied and all Omega chromium-plated nickel cases go through this improved process and are therefore the most reliable that can be produced. Such process has proved satisfactory in other industries, particularly in motor-car accessories.

How to clean chromium-plated watches.

Use a soft dry cloth only — no cleaning powders or liquids required

Cal. 8,1

Baguette movement by Golay in Le Sentier. Width 8.10 mm, lever escapment. Covered winding wheels. From 1930.

Cal. 8,1 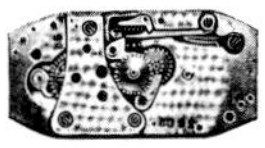   Cal. 8,1

Cal. 37,6

Circular lever movement with 15 jewels for pocket watches. (150) Introduced 1930. Can be found in some large aviator's watches for the wrist. In 1950 it became Caliber No. 150.

150
15 pierres *Jewels*

Cal. 28,9 CHRO
Cal. 33,3 CHRO
(170, 171)

Since 1932 Omega has bought several chronograph calibers from Cal 33,3 CHRO Lemania. While Caliber 28,9 CHRO 1941 was replaced by Cal. 27 (170, 171) CHRO, Caliber 33,3 CHRO remained in the program for many decades. After 1949 it was named Caliber No. 170 (height 6.50 mm, switching wheel) and 171 (water-resistant). From 1935 on the mechanism operated with two buttons.

170   171
17 pierres *Jewels*   17 pierres *Jewels*

153

Cal. 38,5

Circular pocket watch movement, diameter 38.50 mm, height 5.15 mm, 15 jewels. This is an economical movement, which had to be developed for the overseas market due to unfavorable currency exchange rate. It soon became the most popular caliber worldwide and can be also found in oversized observation watches.

160
15 pierres *Jewels*

Cal. 12,6

Tonneau lever movement, 15 jewels, regulated in three positions.

Cal. T 12,6

Tonneau lever movement, 12.60 x 21.90 mm, movement height 3.40 mm, 16 jewels. Introduction in 1934.

Cal. 12,6    Cal. T. 12,6   Cal. T. 17

Cal. T 17

Tonneau Lever movement, 17 x 24.50 mm, height 3.85 mm, 15 jewels, running reserve 60 hours (this is the movement with the longest running reserve), partially set in soft steel cover (antimagnetic). After 1934 this was a very well-liked manual winding caliber. See color illustration page 47.

Cal. 23,4 (220)   Circular lever movement, diameter 23.40 mm, height 3.75 mm, 15 jewels, available since 1935. From 1949 on it was named Caliber No. 220. This Caliber was also available after 1935 with indirect center seconds (movement height 4.50 mm), one example is for the doctor's watch, after 1949 No. 230. Several modifications. Caliber 231 was antimagnetic.

220
15 pierres *Jewels*

230
15 pierres *Jewels*

231
15 pierres *Jewels*

Cal. R 11,5 (210)   Small form movement with lever escapment. 11.50 15 mm, height 3.55 mm, 17 jewels. Available since 1936, was renamed No. 210 in 1949; base caliber. Further alterations were No. 211 (antimagnetic), No. 212 and 213.

210
17 pierres *Jewels*

211
17 pierres *Jewels*

212
17 pierres *Jewels*

213
17 pierres *Jewels*

Cal. R 13,5 (240)   Similar construction as R 11,5; 13.5 x 17.50 mm, height 3.25 mm, 15 jewels. Since 1938, was renamed No. 240 in 1949. Base caliber. Based on this model were Caliber 241 (antimagnetic), 242 (17 jewels), 243 (17 jewels and antimagnetic), 244 (shock resistant), 250 and 251 (center seconds, height 5.10 mm), 252 (center seconds, shock resistance).
Caliber 245 from 1960: 13.50 x 18 mm, height 3.25 mm. Because of the hole for the winding shaft, the movement plate was enforced, yet the thickness of the two bridges was lowered, "soft going" hand setting, lever without stop pins. 17 jewels, four

155

armed glucydur balance, 19,800 half oscillations per hour; breakage and rust resistant spring with six evolutions, running reserve 45 hours.

240
15 pierres *Jewels*

241
15 pierres *Jewels*

242
17 pierres *Jewels*

243
17 pierres *Jewels*

244
17 pierres *Jewels*

245
17 pierres *Jewels*

250
17 pierres *Jewels*

251
17 pierres *Jewels*

252
17 pierres *Jewels*

Cal. R 13,5 SC (250) Similar to R 13,5, but with 17 jewels and sweep seconds, height 5.10 mm. Since 1938; after 1949 renamed No. 250 and 251, No. 252 (with shock resistance).

Cal. 30 mm (260) Robust lever movement, which was well probed; diameter 30 mm, movement height 4.01 mm, 18,000 half oscillations per hour, 15 jewels. It was introduced in 1939, and after 1943 the same wrist-watch movement was produced with shock resistance. The Caliber 30mm was renamed Caliber No. 260 in 1949. Similar Caliber No. 261 (anti-magnetic), No. 262 (special feature for regulation); until 1963 the model was modified six times. The last Caliber number was 269.

Cal. 30 SC (280)

Similar to Cal. 30 mm, but with sweep seconds and 16 or 17 jewels. The 30 T2 SC RG set numerous records at observatory contests. The earlier chronometer wristwatches by Omega had this Caliber. See color illustration page 61. In 1949 it was renamed No. 280 and 281, the latter with RG. The last modification of the Caliber was in 1963 with No. 286 (T6). The compensation balance with Breguet spring had been replaced by a four arm glucydur balance with flat spring. The frequency remained the same with 18,000 half oscillation per hour. The spring, with over seven evolutions, guaranteed a running reserve of at least 42 hours.

260
15 pierres   *Jewels*

261
17 pierres   *Jewels*

262
17 pierres   *Jewels*

265
15 pierres   *Jewels*

266
17 pierres   *Jewels*

267
17 pierres   *Jewels*

268
17 pierres   *Jewels*

269
17 pierres   *Jewels*

280
17 pierres   *Jewels*

281
17 pierres   *Jewels*

157

283
17 pierres *Jewels*

284
17 pierres *Jewels*

285
17 pierres *Jewels*

286
17 pierres *Jewels*

CK 2410 30 SC RG acier inoxydable . . . . Fr. 206.—
CO 2410 30 SC RG acier inoxydable, lunette or 14 ct. . . . . . „ 240.—

⌀ *35 mm, ouverture de lunette 30 mm, cuir 18 mm, cadran 7073 radium*

Majoration pour Bulletin officiel Fr. 12.—

Water-resistant chronometer wristwatch 30 SC RG from 1946.

CK 2364 30 RG acier inox., étui souple . . . Fr. 180. —
CO 2364 30 RG acier inox., lunette or 14 ct.,
                     étui souple „ 225. —
OJ 2364 30 RG or 14 ct., écrin de luxe . . . „ 429. —
OT 2364 30 RG or 18 ct., écrin de luxe . . . „ 497. —

⌀ 33 mm, lunette 29 mm, cuir 17 mm, cadran 4038 heures émaillées
Fr. 15. —

CK 2365 30 SC RG acier inoxydable, étui souple . Fr. 188. —
CO 2365 30 SC RG acier inoxydable, lunette or
           14 ct., étui souple . . . . „ 244. —
OJ 2365 30 SC RG or 14 ct., écrin de luxe . . „ 453. —
OT 2365 30 SC RG or 18 ct., écrin de luxe . . „ 530. —

⌀ 33 mm, lunette 29 mm, cuir 17 mm, cadran 7067
heures or rivées, aiguilles or.   *Fr. 96. —*

Chronometer wristwatch 30 RG and 30 SC RG form the Omega main cata-
logue in 1946. The chronometer certificate increased the price by 12 Swiss
Francs.

159

Cal. R 17,8 (300)

Form movement with lateral positioned escapement, 17.80 x 22 mm, height 3.25 mm, subsidiary seconds, 17 jewels. Introduced in 1939, the model with sweep seconds was introduced in 1941. This Caliber was also used in circular cases. Color illustration page 91.

In 1949 renamed No. 300 and 301; No. 302 has shock resistance, 21,600 half oscillations per hour. No. 310 has sweep seconds, height 4.00 mm.

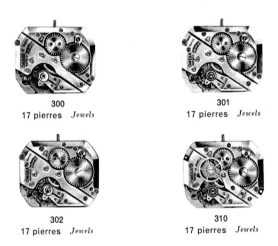

300
17 pierres *Jewels*

301
17 pierres *Jewels*

302
17 pierres *Jewels*

310
17 pierres *Jewels*

Cal. 27 CHROA (320)

Small caliber (diameter 27.00 mm, height 5.57 mm, 18,000 half oscillations per hour, 17 jewels), was added to the chronograph program in 1941. After 1949 No. 320. No. 321 (height 6.74 mm, 12 hour register, longer index as protection for hair spring).

320
17 pierres *Jewels*

321
17 pierres *Jewels*

Cal. 30,10 RA (330) Thin self-winding lever movement, diameter 30.10 mm, height 4.30 mm, 17 jewels. This is Omega's first self-winding gentlemen's movement, which was added to the production in 1942. The oscillating mass, set in jewels, was not yet rotating but moving only back and forth, while the springs had to absorb the force of the impact. However, the pressure was so strong on screws that they became loose and the buffers screws sometimes broke. In order to minimalize this problem, the buffers received small feet which were anchored in the movement plate. Color illustration page 118. When continuous Caliber numbers were introduced in 1949, the self-winding movement received Caliber No. 330 and the one for the Centenary Model in 1947 the No. 331 (shock resistance, antimagnetic, 19,800 half oscillations per hour). Other modifications were No. 332 and No. 333 (with the special construction of Caliber 30 RG).

330
17 pierres *Jewels*

331
17 pierres *Jewels*

332
17 pierres *Jewels*

333
17 pierres *Jewels*

Cal. 28,10 RA (340)   The Caliber 30,10 RA was also offered in a slightly smaller version (diameter 28.10 mm), but with a larger height (4.80 mm). This caliber was used for the Seamaster in 1948 and the Constellation in 1952. Winding with ratchet.

After 1949 it was known as No. 340, 341 (JUB), 342, 343 (RG), 344 (swan's neck regulation). With sweep seconds it was named No. 350, 351, 352 (RG, height 5.40 mm, 19,800 half oscillations), 353 (date), 354 (swan's neck regulation), 355 (swan's neck regulation and date, height 6.10 mm).

Caliber 351 was also available with a subsidiary dial in the center for winding indication.

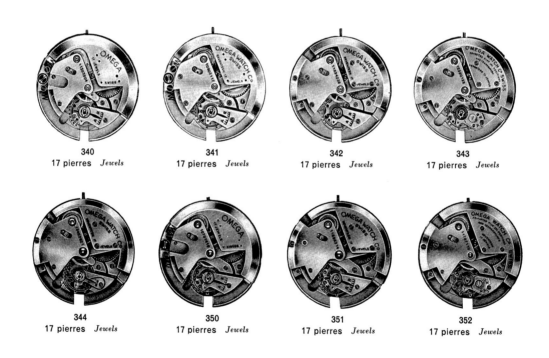

340	341	342	343
17 pierres *Jewels*	17 pierres *Jewels*	17 pierres *Jewels*	17 pierres *Jewels*

344	350	351	352
17 pierres *Jewels*	17 pierres *Jewels*	17 pierres *Jewels*	17 pierres *Jewels*

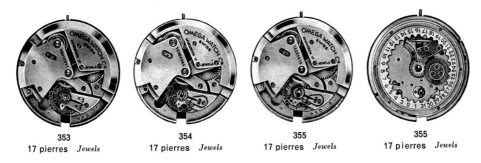

353
17 pierres *Jewels*

354
17 pierres *Jewels*

355
17 pierres *Jewels*

355
17 pierres *Jewels*

Cal. 28 (360)

Thin lever movement with manual winding, diameter 28.00 mm, height 3.25 mm, 18,000 half oscillations per hour, 17 jewels, subsidiary or sweep seconds. Similar to the Caliber 30 mm, but thinner. Was introduced to the market in 1944. After 1949 listed as No. 360, 361 (PC), 370 (SC) and 371 (SC, PC).

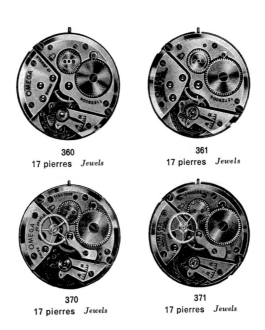

360
17 pierres *Jewels*

361
17 pierres *Jewels*

370
17 pierres *Jewels*

371
17 pierres *Jewels*

Tourbillon

Circular Precision lever movement with 3/4 plate and asymmetric pierced nickel-copper alloy bridge, damascent, diameter 30 mm, the 7 1/2 minute tourbillon base is silvered brass, bimetallic Guillaume balance with 18 adjustment screws,

163

spring with Philipscher end curve, 18,000 half oscillations per hour, 23 jewels, subsidiary seconds, hands set via pin at 4 o'clock. manual winding. Specially made for participation in timing contests since 1947. The Tourbillon was made to adjust for the effect of gravity especially in vertical position. Color illustration page 60.

Cal. 27 DL (381)   Circular lever movement with full calendar and phases of the moon indication, diameter 27 mm, height 5.25 mm, 18,000 half oscillations per hour, 17 jewels, shock resistant. For the model Cosmic, which was brought into the collection in 1947. For a further discussion on the calendar mechanism see page 165 and 166. Except for the phases of the moon, all indications are set automatic. Corrections are made over setting pins on case band. After 1949 named Caliber No. 381.

Cal. 410   26,5 PC AM 17p; height 3.60 mm, 18,000 half oscillations per hour. 1951. Cal. 420 (SC), Market introduction in 1952.

410
17 pierres *Jewels*

420
17 pierres *Jewels*

Cal. 440   13,5 PC AM 17P, circular lever movement with manual winding, height 3.15 mm, 21.600 half oscillations per hour, running reserve 40 hours, covered winding, 1954.

440
17 pierres *Jewels*

Calibre 381 (27 DL PC AM 17 pierres)

D: 27.00 mm
Ht: 5.25 mm
N: 18000

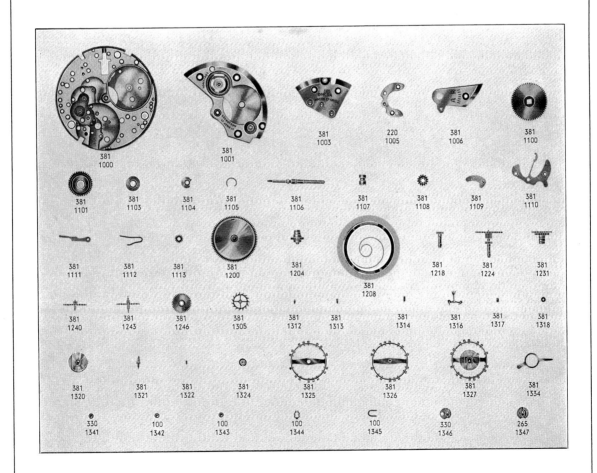

Calibre 381 (27 DL PC AM 17 pierres)

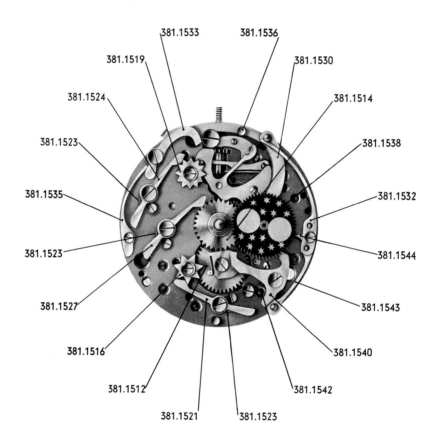

381.1512	Calendar cam wheel		381.1524	Jumper for month
381.1514	Calendar wheel		381.1523	Spring for jumper
381.1516	Day of the week dial star wheel		381.1527	Jumper for calendar wheel
381.1519	Month dial star wheel		381.1523	Spring for jumper
381.1521	Jumper for day of the week		381.1530	Date change lever
381.1523	Spring for jumper		381.1532	Spring for date change lever
			381.1533	Month change lever

Cal. 372

28 SCS 17P, circular lever movement in bridge layout, Incabloc shock resistance, special index regulator, 18,000 half oscillations per hour, manual winding. A special mechanism makes the sweep seconds jump after five half oscillations by one second (first jump second mechanism by Omega). 1952.

372
17 pierres *Jewels*

Cal. 480

12,5 PC AM 17P, tonneau movement, 12.50 x 15.20 mm, height 3.60 mm, wavy pink gilding, diamond polished edges, side anchor in Duofix setting, balance and spring are antimagnetic, 19,800 half oscillations per hour, large spring barrel, running reserve up to 40 hours. 1955.

Modifications of this Caliber are No. 481, 482, 483 (without Duofix setting, since this seemed to create difficulties with cleaning and oiling the movement), 484 and 485.

Caliber 482 received in 1959 a plain glucydur balance.

480
17 pierres *Jewels*

481
17 pierres *Jewels*

482
17 pierres
Jewels

483
17 pierres
Jewels

484
17 pierres
Jewels

167

Movement 480, which was added to the caliber production in 1955.

Cal. 455

First Omega Caliber with Rotor winding, diameter only 16.50 mm, SC, PC (KIF), AM, 17P, height 5.50 mm, 19,800 half oscillations per hour, automatic movement with rocking arm, third wheel with transmission wheel on pivot, which is connected to the center seconds wheel, beryllium bronze spring powers the running gear. 1955. By 1961 208,000 pieces of this ladymatic movement had been made, which originally were sold with a chronometer certificate. Enlarged illustration of movement on page 171.

Cal. 470

25 RA SC PC AM 17P. Automatic Caliber with rotor winding in both directions, height 5.50 mm, central seconds, 19,800 half oscillations per hour. Micrometer adjustment to index. Market introduction 1955. Enlarged illustration on page 172 and 173.
Modified Caliber 471 (19 or 20 jewels).

470
17 pierres *Jewels*

471
19 pierres *Jewels*

471
20 pierres *Jewels*

490
17 pierres *Jewels*

Cal. 490

28 RA PC AM 17P, rotor automatic, 19,800 half oscillations per hour, swan's neck fine regulation, subsidiary seconds. Market introduction 1956. Related caliber 491 (19P). This movement can also be found with regulator dial.

169

Cal. 500

Similar to Caliber 490 but with additional center seconds, made since 1956.
Caliber 500 (17P), 501 (19 or 20 P), 502 (17 P, CAL), 503 (19 or 20P), CAL with very simple mechanism and low height), swan's neck regulation, 504 (24 P, for chronometer) and 505 (24P, without calendar).

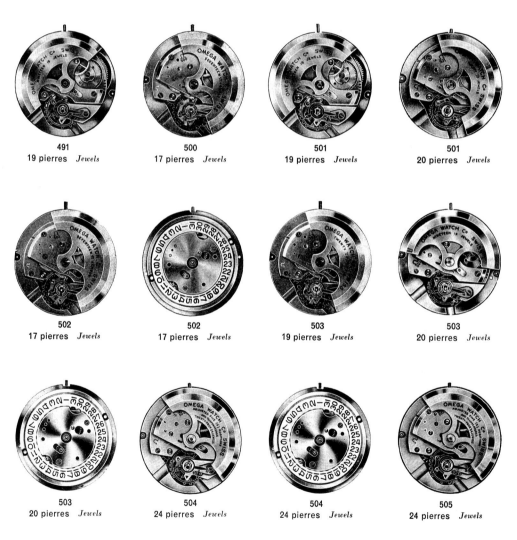

491	500	501	501
19 pierres *Jewels*	17 pierres *Jewels*	19 pierres *Jewels*	20 pierres *Jewels*

502	502	503	503
17 pierres *Jewels*	17 pierres *Jewels*	19 pierres *Jewels*	20 pierres *Jewels*

503	504	504	505
20 pierres *Jewels*	24 pierres *Jewels*	24 pierres *Jewels*	24 pierres *Jewels*

The simple mechanism for the automatic winding of Caliber 455 was patented.
All parts are set on one block which consists of two bridges, which are held
together by two screws and two locating pins.

Caliber 471. When the winding rotor turns to the right, wheel A connects directly to the changer wheel R, which transmits power directly to wheel M; if it turns to the left, the changer wheel R shifts and the toothed wheel R1 is set in between. Reduction wheel M then transmits the energy to ratchet wheel and mainspring. The click mechanism C, activated by F, holds wheel M and hinders the wheels from running backwards. Caliber description page 169.

Cal. 510   25 PC AM 17P. Flat and circular lever movement with 3/4 plate and manual winding. Diameter 25.50 mm. height 3.25 mm, 19,800 half oscillations per hour, subsidiary seconds, same movement layout as automatic caliber 471, spring with six evolutions, running reserve 40 hours. 1956. Enlarged movement illustration on page 175.
Caliber 511 without subsidiary seconds.

Cal. 520   Was a modification of Caliber 510 and taken into the program line in 1957. Diameter 25.50 mm, height 3.80 mm, glucydur balance with 18 setting screws, sweep seconds, swan's neck fine regulation, adjusted to two positions, running reserve 40 hours. manual winding. Enlarged movement illustration on page 176.

Cal. 540   20,50 AM 17P. Ultra thin and rather small caliber, height 2.00 mm, one bridge for spring barrel, interim wheel and minute wheel; one bridge for seconds, seconds and escape wheel; wavy pink gilding and diamond polished edges; 17 jewels, 12 of which are hole jewels set in the running gears and escapement, two pallet jewels, two balance cap stones, one endstone; balance wheel and spring constitute an antimagnetic unity (beryllium bronze balance wheel with screws, compensating flat spring, the girocapchaton allows a precise setting of the height of balance wheel, thus allowing for precise oiling and lubrication, thus extending the servicing intervals; 19,800 half oscillations per hour, the hand setting mechanism has a second gear connecting to the changing wheel; regulator without index, a groove at the regulator ring and an engraved scale on the balance bridge facilitate the reading of regulation; adjusted to two positions, running reserve 38 hours. Market introduction in 1957, movement illustration on pages 177 and 178.

Manual winding Caliber 510 from 1956. Diameter 25.50 mm. Description on page 174.

Manual winding Caliber 520 from 1957, enlarged. Diameter 25.50 mm. Description on page 174.

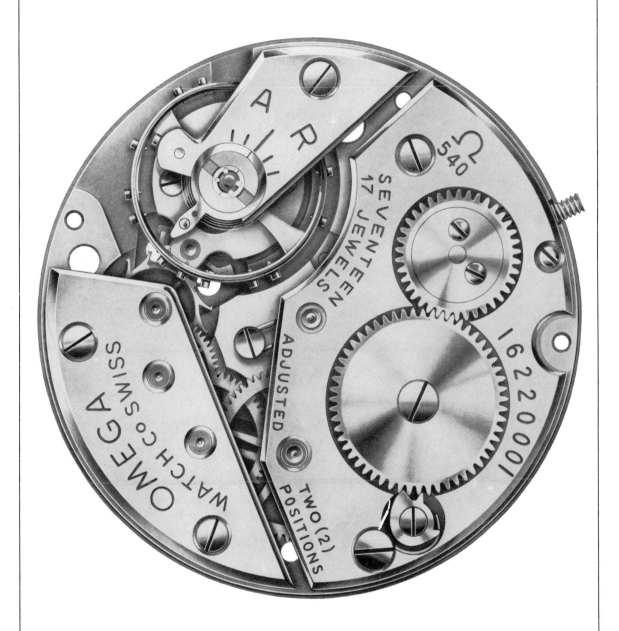

Manual winding Caliber 540 from 1957, enlarged. Description on page 174.

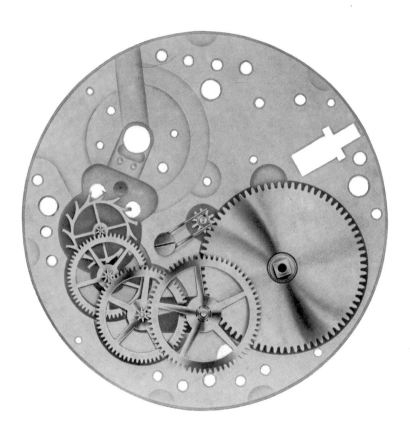

Caliber 540 has a layout which differs very much from the usual watch. Due to the thinness of the movement, the minute wheel was replaced by an interim wheel. While the spring barrel drives the minute gear, the interim wheel connects to the second wheel.

Cal. 550

27,9 RA SC PC AM 17P. New, ultra thin automatic movement from 1959, height 4.50 mm. Base Caliber of a very famous Caliber generation. It replaced the rotor automatic 470 and 490, which showed wear to the wheel which had a spring driven click.

Cal. 551

Similar to caliber 550, 24 jewels and chronometer certificate, made exclusively for the Constellation. Market introduction in 1959. Movement illustration very enlarged in pages 180 and 181.
Winding activated by the smallest movement by central rotor in both directions, plain glucydur balance, 19,800 half oscillations per hour, swan's neck fine regulation, first flexible spring block for easier regulation, adjusted to five positions, running reserve over 50 hours (the spring has seven evolutions).
Additional calibers related to Caliber 550 are Caliber 552 (similar to 551, but without buttelin), 560 (17P, CAL) for the Seamaster de Ville, 561 (24P, CAL, BULL), 562 (24P, CAL), 563 (17P, CAL, CORR = Calendar quick change), 564 (24P, CAL, CORR, BULL) and 565 (24P, CAL, CORR). Pages 182/183.

Cal. 570

24,9 RA SC PC 17. Height 4.50mm, 19,800 half oscillations per hour, swan's neck regulation, adjusted to two positions. Market introduction in 1959. Related Caliber is Caliber 571 (24P).

570
17 pierres *Jewels*

571
24 pierres *Jewels*

The automatic chronometer Caliber 551, which was inside the Constellation from 1959 on. Description on page 179.

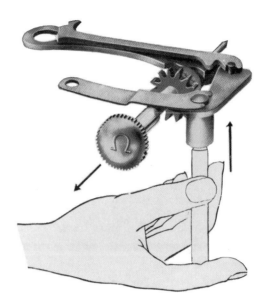

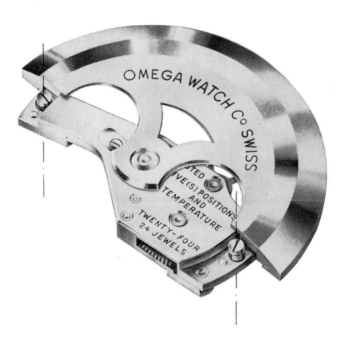

The automatic winding mechanism of Caliber 551 is attached to the movement with two screws. In order to remove the movement from the case it is sufficient to push a pin against the upper end of the cam of the setting lever to remove the winding wheel. The setting lever screw is replaced by a cam connected by a setting lever. A plate spring presses the setting lever against the movement plate.

181

The chronometer Caliber 561 with rotor winding and date.
Description on page 179.
The calendar mechanism works as follows: The hour wheel is connected to a twelve toothed-wheel, which has a second wheel with eight teeth attached, which in turn is connected to a date indication wheel (48 teeth). The latter turns once every 24 hours. The date change starts at 22 o'clock and is completed at 24 o'clock.

The Calendar mechanism of the Chronometer Caliber 564. When the crown is in position "3," a spring pushes the date change lever into the teeth of the date wheel, turning it by one tooth. This procedure is to be continued as long as necessary.

Cal. 580	13,50 REM PC AM 17P. Circular lever movement from 1959. similar to Caliber 440; height 3.20 mm, 21,600 half oscillations per hour, adjusted to two positions. Enlarged movement illustration in page 186.
Cal. 591	27,8 RA SC PC AM 20P. Introduction 1960. Simple manufacture, rational in large series. Height 5.35 mm, flexible spring block, 19,800 half oscillations per hour, running reserve 48 hours. Enlarged movement illustration on pages 187 and 188.
Cal. 600	27,9 SC PC AM 17P REM. Swan's neck fine regulation, height 3.85 mm, 19,800 half oscillations per hour, running reserve 50 hours. 1960. Enlarged movement illustration on page 189. Related Calibers 601 (different regulator), 692 (BULL, different regulation), 610 (CAL), 611 and 613 (quick date change)
Cal. 620	17,5 PC AM 17P. Balance will with drill holes for equilibrium adjustment, height 2.50 mm, 19,800 half oscillations per hour. 1962. Enlarged movement illustration on page 190. Related Calibers 627 (DL), 630(SC), and 635 (SC, T1). Caliber 626 (17,5 PC, 17P), 1984 for Louis Brandt Collection.
Cal. 690	R 7 REM PC AM 17P. Baguette movement 7.00 x 18.80 mm, height 3.78 mm, waverly pink gilding, lever with border pins, KIF-shock resistance, 21.600 half oscillations per hour, running reserve 42 hours, crown for winding and setting on back, regulator without index. 1962. Enlarged movement illustration on pages 197 and 198.
Cal. 640	12,4 PC AM 17P REM. Height 2.85 mm, lateral layout for anchor escapement (anchor, anchor wheel and balance not in one line), instead of border pins there

are cuts on the plate, 19,800 half-oscillations per hour; open spring barrel, back wind (crown can be turned only in one direction, a disconnection is not possible). For lady's and jewelry watches. Market introduction in 1963. It replaced Caliber 440. Enlarged movement illustration on page 191.
Related Calibers 650 (with traditional manual winding on the side of the case), it replaced caliber 580. Enlarged movement illustration on page 192.

Cal. 661   15 RA PC AM 24P. Lady's Caliber with rotor winding (automatic with 7 jewels), diameter 15.40 mm, height 4.25 mm, flexible spring block, 19,800 half oscillations per hour, adjusted to two positions, running reserve 40 hours. 1963. Enlarged movement illustration on pages 193 and 194.
Base Caliber 660 with 17P. Related Calibers 662 (17P, T1) and 663 (24P, T1).

Cal. 670   17,5 RA SC AM 17P. Height 4.30 mm, 19,800 half oscillations per hour. Simple automatic caliber for larger lady's and thin gentlemen's models. 1963. Enlarged movement illustration on pages 195 and 196.
Alterations of this Caliber are 671 (24P), 672 (24P, BULL), 680 (17P, CAL), 681 (24P, CAL), 682 (24P, CAL, BULL), 683 (17P, T1), 684 (24P, CAL, T1) and 685 (24P, CAL, BULL, T1).

Cal. 686   20RA PC CAL 24P. 1984 for Louis Brandt Collection.

Cal. 688   20 RA PC 24P. 1984 for Louis Brandt Collection.

Cal. 700   Ultra thin circular lever movement with manual winding, diameter 20.40 mm, height 1.76 mm, open spring barrel, glucydur balance with 14 gold screws, ultra thin compensation spring, 17 jewels, limited space between diamond hands, 18,000 half

185

The very small lady's Caliber 580, whose diameter is only 13.80 mm. Description on page 184.

*Caliber 591 from 1960, which is a very simple gentlemen's self-winding
Caliber. Description on page 184.*

The self-winding mechanism of Caliber 591 is connected to the movement
with only three screws on the spring barrel bridge.
The third wheel runs above the minute wheel and the spring barrel bridge. It is
also connected to the seconds wheel, which makes the transmission wheel
obsolete. Movement numbers 17420000 to 1743999 are numbered on the
top rotor bridge, movements starting at 17640000 are numbered on a movement bridge.

The fine manual winding Caliber 610 from 1960. Description on page 184.

Caliber 620 with manual winding from 1962 for lady's watches (diameter 17.50 mm). Description on page 184.

The small lady's Caliber 640 with back wind and diameter 12.40 mm. Descrip-
tion on page 184.

Lady's Caliber 650 from 1963 with diameter 12.40 mm. Description on page 185.

Lady's Caliber 661 with rotor automatic from 1963. Description on page 185.

One of the characteristics of Caliber 661 is the unusual construction of the
crownwheel.

The movement of the crownwheel is transmitted to the barrel rachet wheel via two
gears. Those two gears are called the crownwheel gear and the reversing gear.
The winding bridge, which is attached to the barrel bridge with two screws and two
setting pins, is the base for this winding gears. Both the barrel bridge and the crown
wheel bridge have an oval cut out, in which the pinions of the reversing gears are
set. The automatic winding mechanism is sitting on the movement plate with two
screws and two setting pins. This layout allows for a small height of the movement
without having an impact on the movement gears.

The outer ring of the rotor is heavy metal. The rotor gear is a beryllium and bronze
alloy, because of its low friction coefficient. The smallest movement of the wrist
moves the rotor and guarantees a perfect winding action.

Simple automatic Caliber 671 with diameter 17.50 mm and a height of 4.30 mm. Description on page 185.

195

One of the characteristics of Calibers 671 and 681 is the unusual crown wheel construction. The movement of the crown wheel is transmitted to the barrel rachet wheel via two gears. The crown wheel bridge, which is attached to the barrel bridge with two screws and a setting pin and the starting point for the crown wheel gear, is the mounting for these gears.

Baguette Caliber 690 (7.00 x 18.80 mm) with back wind from 1962.
Description on page 184.

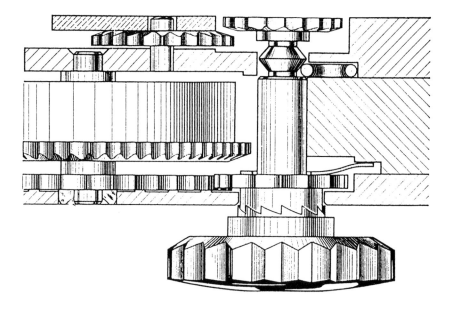

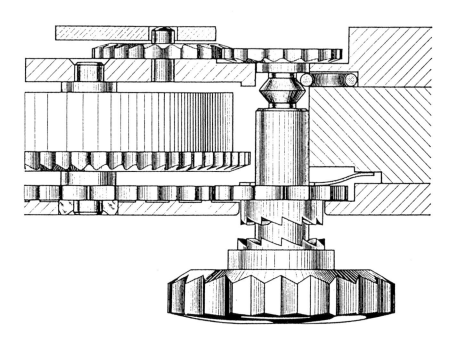

198   *The back winding and hand setting mechanism for baguette caliber 690.*

oscillations per hour, adjusted to two positions, base caliber by Piguet (9ligne). It replaced Caliber 540, which was approximately 0.30 mm higher. Market introduction in 1964. Enlarged movement illustration on page 201.

Cal. 710 25 RA SC PC 17P. Rotor automatic, height 3.00 mm, off-center fine regulation, hands mechanism with flexible gear, 19,800 half oscillations per hour, adjusted to two positions, running reserve 43 hours. 1966. Enlarged movement illustration on pages 202 and 203.
Related Calibers are 711 (24P), 712 (24P, BULL) for Constellation, 715 (24P), 712 (24P, DL, CALP) 1984 for Brandt Collection and 717 (24P, CAL, FU) for Brandt Collection.

Cal. 720 17,5 RA SC CAL, Base Caliber ETA 2681. 1989

Cal. 721 19,4 RA DL. Base Caliber ETA 2685, 1990.

Cal. 730 R 9 PC 17P. Lady's Caliber 9.00 x 16.40 mm, height 3.20 mm, very small movement with large oscillating weight, side anchor, 21,600 half oscillations per hour, running reserve 48 hours. Market introduction in 1968. Enlarged movement illustration on page 204.

Cal. 750 27,9 RA SC CALD CORR AM 17P. First Omega Caliber with date and day of the week, swan's neck fine regulation. 1967.
Related Calibers 751 (24P, BULL) and 752 (24P, without BULL)

750
17 pierres *Jewels*

751
24 pierres *Jewels*

752
24 pierres *Jewels*

Cal. 860   27 CHRO; 17 jewels, 21.600 half oscillations per hour, micrometer regulator with eccenter screw, running reserve 50 hours, manual winding, simple yet robust chronograph mechanism without column wheel and 30 minute register; enlarged movement illustration on pages 205 to 209. 1969. Related Calibers 861 (12 hour register, antimagnetic cover) for Speedmaster Professional Mark II, 862 (21P), 863, 865 (Chronostop) and 866 (17P, C12, DL).

Cal. 910   Similar to Caliber 860, with additional AM/PM indication and dual time zone. For flightmaster. 1969. Caliber 911 (C12, GMT), 920 (Chronostop with CAL), 930 (Chronograph with CAL).

Cal. 980   30,8 RA SCD PC CORR 19P REV; Lemania-Ebauche. Fine regulator with eccenter screw, 21.600 half oscillations per hour, height 7.80 mm, quick date change by pin on case, Memomatic. 1969. Alarm: hammer on short sound spring.
This wrist alarm was described in the DUZ 5/72 in great detail, in order to educate watchmakers about possible repairs. Because the construction details are of importance to collectors and watchmakers alike, they are described below:
"The movement has a central rotor and only one barrel for both the going and alarm train, similar to a traveling alarm with one-key winding. The small locking wheel has a pin at the bottom side, which is corresponding to an index on the spring barrel, which determines that the alarm gear is running for only one evolution of the spring at a time. In order to release the spring, the locking wheel has to be removed, and the spring will release its energy via the alarm gear, after, of course setting the alarm and hands. The automatic winding mechanism can be removed en bloc after removing two

The ultra thin Caliber 700, with height 1.76 mm. Description on page 185.

Caliber 711 from 1966, an ultra thin automatic Caliber with rotor winding in both directions. Description on page 199.

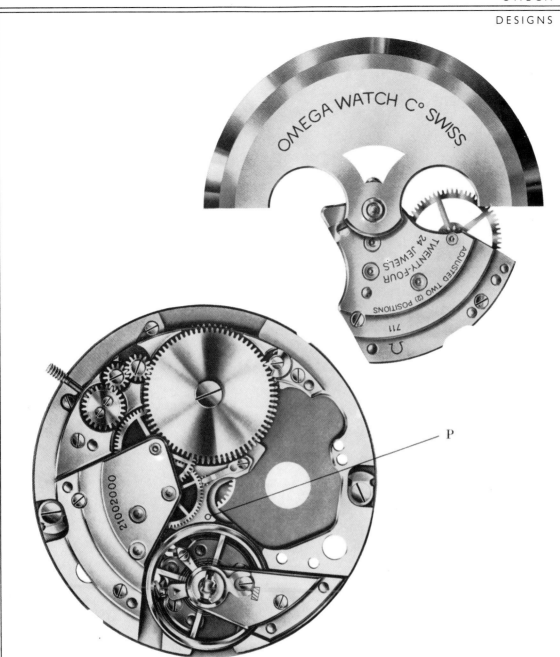

P

The first series of Caliber 711 movements have a pin on the automatic winding mechanism which corresponds to drill hole P on the movement plate. Subsequent series have a pillar on the plate instead. The pin in movement numbers 21,000,000 to 21,006,999 is 0.2 mm higher than with the following movements. This will mean a change to the automatic winding bridge for movement from Number 21,007,000 on. If the automatic winding bridge has to be exchanged for movement number 21,000,000 to 21,006,999, it is necessary to shorten the pin by 0.20 mm. The rotor winding in both directions is activated by the slightest movement and guarantees an above average winding action.

204   *Caliber 730 from 1968. Description on page 199.*

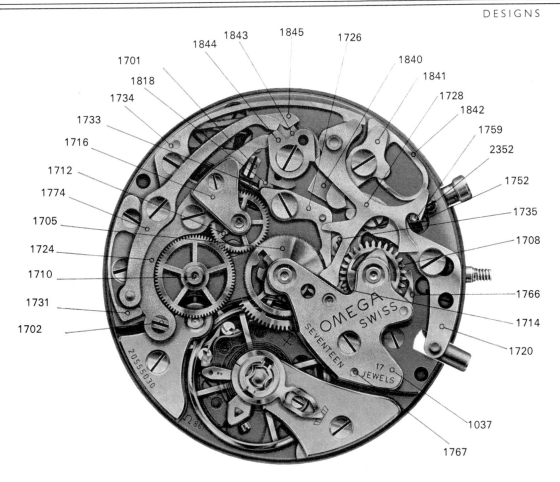

1037	Chronograph bridge	
1701	Eccentric pin for intermediate chronograph wheel lever	
1702	Eccentric pin for intermediate chronograph wheel lever	
1705	Center chronograph wheel	
1708	Minute count wheel	
1710	Fourth wheel	
1712	Intermediate chronograph wheel	
1714	Intermediate minute counter wheel	
1716	Intermediate chronograph wheel bridge	
1720	Start lever	
1724	Intermediate chronograph wheel lever	
1726	Locking lever	
1728	Heart lever	
1731	Intermediate chronograph wheel spring	

1733	Locking lever spring
1734	Heart lever spring
1735	Friction spring for center chronograph wheel
1752	Spring for heart lever lock
1759	Heart lever lock
1766	Minute register wheel click
1767	Spring for minute count wheel click
1774	Hour register wheel lever
1818	Locking lever
1840	Connecting lever
1841	Switch lever
1842	Switch lever spring
1843	Switch disc, bottom
1844	Switch disc, top
1845	Switch disc lever
2352	Screw for return to zero push shaft

Parts for chronograph mechanism Caliber 860, 861, 910, 930. See also page 200.

205

Caliber 865 from 1967, lever movement "Chronostop" for young clients.
Simple and robust chronograph mechanism (Stop and return to zero mecha-
nism with one button), leather strap with buckle. Models "Geneve" and
"Seamaster." See also page 200.

206

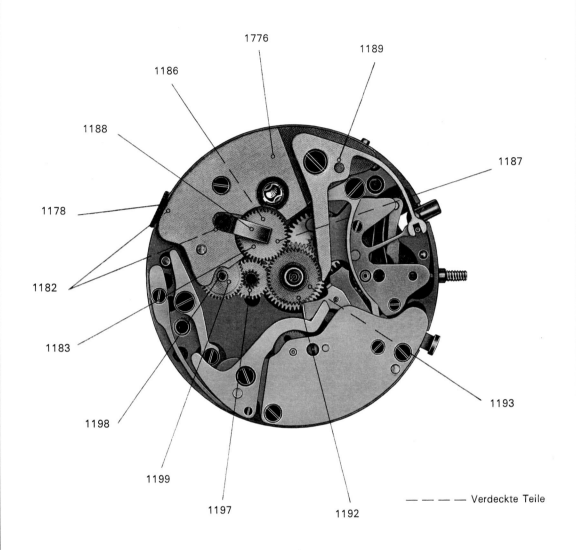

1776
1189
1186
1188
1187
1178
1182
1183
1198
1193
1199
1197
1192

— — — — Verdeckte Teile

1178 Setting crown for GMT
1182 Setting wheel I and II for GMT (2
    wheels)
1183 Setting wheel III for GMT
1186 Pawl for GMT
1187 Pawl spring for GMT
1188 Bow for setting gear III GMT
1189 Stop lever for GMT

1192 Hand wheel for GMT
1193 Friction spring for hand wheel for
    GMT
1197 Interim wheel AM/PM
1198 Pinion for hand wheel AM/PM
1199 Hand wheel AM/PM
1776 Underdial plate

The AM/PM and GMT mechanism for Caliber 910. See also page 200.

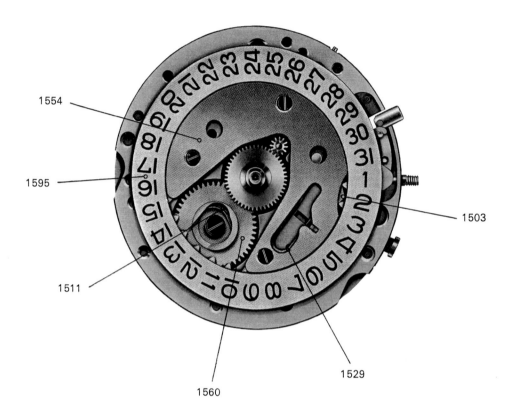

1503 Date lock   1554 Under date chapter plate
1511 Switch index   1560 Date chapter wheel
1529 Date lock spring   1595 Date chapter

The Calendar mechanism for Caliber 930. See also page 200.

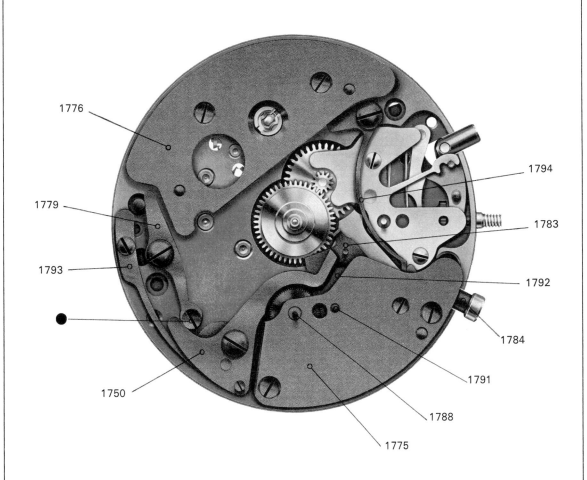

1776

1779

1793

1794

1783

1792

1784

1791

1788

1750

1775

1750   Hour register stop lever
1775   Hour register wheel bridge
1776   Under dial plate
1779   Disconnecter lever
1783   Hour register heart lever
1784   Push for hour register heart lever
1788   Hour register wheel

1791   Running gear for hour register wheel
1792   Friction spring for running gear (hour register)
1793   Hour register stop lever spring
1794   Hour register heart lever spring
●   Eccentric pin (Mounted on part 1779)

The hour register mechanism for Calibers 861 and 910. See also page 200.

209

screws and the alarm mechanism is also set on its own plate. After removal of the locking wheel (to release the spring) and the date chapter ring (to avoid damage to its lacquer surface), the remaining movement can be cleaned fully assembled employing the vibration technique. However, the alarm wheel is made of plastic and the use of dry lubricants should be done carefully, because these products may damage or destroy certain plastics. After these references the author focuses on the alarm mechanism itself and the difficulties the watchmaker might face during his work:

"So far, the movement plate side is rather problem-free, while the dial plate side can be tricky. There are several points, which, although not obvious at first, can turn into problems. One might get very frustrated when trying to coordinate alarm setting, hand setting and date; however, once one knows how to do it, it is actually very simple.

The dial (or chapter ring, respectively) is not, as usual, attached by pillar and screw, but rather by pressure. A groove on the dial/chapter ring and a pin on the dial plate have to be aligned and give the right position. After the chapter ring is lifted, the alarm setting mechanism (i.e., the hour chapter ring and minute disc) can be removed. Underneath the other mechanism parts such as hands gears, date mechanism, alarm release, et al."

The instruction booklet further continues: "After removal of the alarm setting unit, the toothed wheel lays open. It is attached by three pins and can be simply lifted off without loosening any screws. The under plate is screwed. The holding ring for the date chapter is also screwed, and when reassembling one has to pay attention, that the jumper of the date change lever is positioned correctly; the jumper is controlled in its movement by a pin on the holding ring and a false setting of the pin would result in a jamming of the date indication."

When reassembling the movement, hand setting, alarm setting and date had to be coordinated, which was a very demanding task, even for the experienced watchmaker:

"First, it is important to know that the alarm indication mechanism can be taken apart! It consists of a base disc, with a wheel set similar to the hands gear, the minute disc and the hour chapter. The chapter for the minute indication is pressure attached to the base disc and after its removal, the minute disc can also be removed.

The reassembling of the dial side is as follows: the alarm crown has to be set to "alarm." The winding crown has to be pulled and the hands turned until the date changes. At this position the base plate of the alarm mechanism is positioned in correspondence with the three indexes on the hour wheel, which is the alarm position. The minute disc has to be set in a way that the index shows 12 o'clock. Position "twelve" corresponds with the setting pin for the chapter ring. The minute chapter is next, it is also installed in coordinance with 12 o'clock. Then the chapter ring is pressed on only after attaching the thin flexible alarm indication ring. The hour and minute hand are both set at zero and "twelve o'clock" respectively. It is helpful to do this whole operation with released spring or removed balance, in order to avoid that the time continues running and hence endangering the exact accord of date, alarm and time setting."

The brochure closes with the following words: "If all these steps are carefully followed, you should not have any problems repairing a Memomatic."

Cal. 1000   27,9 RA SC CAL CORR STS 17P. 1969. Height 4.00mm. 28,800 half oscillations. Related calibers 1001 (20P, BULL), 1002 (20P), for the Seamaster 600, 1010 (17P), 1011 (23P, BULL), 1012 (23P).

Cal. 1020   Similar to base Caliber 1000, but in addition has CALD CORR CORS 17P. 1973 for Seamaster. Related Calibers 1021 (23P, BULL), 1022 (23P, without BULL).

Cal. 1030	Modified manual Caliber 600 from 1960, in addition CAL and CORR and stop seconds. 1974. 1035 without CAL.
Cal. 1040	31 CHRO C12 RA PC CAL CORR 22P. 1970 for Speedmaster Mark II Automatic. 1041 (BULL) for anniversary Chronograph Omega 125 in 1973. 1045 (CALD, CORR, CorJ, PS, A24, STS, PC, 17P) for Speedmaster Automatic Day-Date.
Cal. 1060	13,5 PC 17P. Prototype. 1974.
Cal. 1070	R 13 PC 17P. 1974.
Cal. 1100	R 13 PC 17P. 1977.
Cal. 1110	25,6 RA SC PC CAL CORR STS 21P. 1984. Base Caliber ETA 2892-2. 1111 (BULL), 1988 for Seamaster Professional 200m, Chronometer. 1112 (DL) 1990, 1113 (DL, CALP). 1990.
Cal. 1140	29,3 CHRO RA C30 C12. Base Caliber ETA 2890-2. 1988 for Speedmaster automatic.
Cal. 1150	29,9 CHRO RA DL CAL MO. Base Caliber TA 7751. 1989.
Cal. 1154	29,9 CHRO RA C30 C12 CAL 17P. Ball bearing rotor, winding in one direction, Incabloc shock resistance, 28,800 half oscillations per hour, chronometer quality. Diver's chronograph 300m. Base Caliber ETA 7750.
Cal. 1480	25,6 RA SCD PC CAL CORR 17P. 21,600 half oscillations per hour. 1481 (21P).

Louis Brandt, fondateur.

Louis-Paul Brandt.

César Brandt.

Adrien Brandt,
président

du Conseil
d'administration.

Paul Brandt,
administrateur.

Ernest Brandt,
administrateur.

Gustave Brandt, administrateur.

The History of Louis Brandt & Freres/Omega AG

1825   The founder of the company, Louis Brandt is born on May 13 in La Brevine.

1848   At the age of 23 Louis Brandt started a company in the watchmaker's village La Chaux-de-Fonds, which made pocket watches from supplied parts.

1850   The owner of the company starts regular trips by stagecoach through Europe with his collection (England, France, Belgium, Netherlands, Scandinavia, Austria, Italy, and Germany, where he participates at the Leipzig Fair).

1877   His Son Louis-Paul becomes a partner in the company.

1879   The founder of the company dies on July 5, at the age of 54 in La-Chaux-de-Fonds.

1880   The two sons, Louis-Paul and Cesar Brandt, move the Comptoir d'etablissage (Avenue Leopold-Robert No. 59) to the little town of Biel about 50 kilometers away. There, on the second floor of the company, Schneider & Perret-Gentil in Bozinger str 119, they start their own manufacturing company. The same year, they purchase the entire building and lot, including the steam engine located in an adjoining building.

1882   In June the watchmakers purchase the weaving company Bloesch-Neuhaus founded in 1825 in nearby Gursele. The large grounds included a multi-story main building with two side wings, a building containing the steam engine, the manor house, three apartment buildings, and a farm and stables. Here a new small watch company is started, and here the Brandts started their watchmaking empire.

1885   Along with cylinder watches, the first pocket watch with lever movement is manufactured, the so-called Labrador. At the World Exhibition in Antwerp, the Brandts exhibited their Leisure pocket watch called Decimal, a thick model, which also had a dial on the back (for example, with

Brandtsche Manufacture in Bozingerstr. (Biel)

215

a playing card arranged to simulate a dial). Pushing a pulled button made the hand jump from its zero position and stop at any field.

1888   Cesar Brandt moves to Paris to better coordinate the international operations of the company.

1889   The Swiss State Council names Cesar Brandt a member of the international jury for the World Exhibition, category watchmaking. This was considered a special distinction at the time, and was considered more important than a first prize for a winning product. The jury member nomination was an acknowledgment of the Brandts' importance in Swiss watchmaking industry. The company had become the largest Swiss company, employing 600 people with a production of about 100,000 watches per year.

1891   The company Louis Brandt & Fils is changed to a collective partnership Louis Brandt & Freres.

1894   Building of a new machine park, which allowed for the manufacture of high quality watches in an industrial setting. The parts were so precisely manufactured that they were interchangeable without manual adjustment. The famous and legendary pocket watch caliber OMEGA (diameter 42.86 mm) is introduced. The owners of the company registered the letter Ω and the word OMEGA as new trademark the same year.

1896   Founding of the separate case manufacturing company La Centrale by the Brandts and the industrialist E. Boillat. The location was within the Omega area in Gurzelen. Omega is awarded a gold medal at the national exhibition in Geneva.

1897   The number of employees has reached 800, the yearly production reaches 200,000 watches. At the international exhibition in Bruxelles, Omega is awarded a Grand Prix.

1898   Omega offers its first chronograph (with sweep seconds hand and 30 minute register).

1900   Participation at the world exhibition in Paris. The company is awarded a Grand Prize for the pocket watch movement Greek Temple. The first time wristwatches are part of the collection.

1903   Death of both founders: in the spring Louis-Paul, and in the fall Cesar is laid to rest. The third generation steps in; the company is transformed into a corporation (Societe anonyme Louis Brandt & Freres—Omega Watch Co)

1905   More and more of Omega's precision watches are sold with chronometer certificates. In 1905 42% of all certificates issued by the Observatory in Geneva are for products of the Bieler Company L. Brandt & Frere.

1906   Large success for Omega at the World Exhibition in Milano: the company is awarded a Grand Prix.
The main catalogue contains 400 different pocket watches, ranging from a simple engine turned model to a casted and enamelled model.

216   1914   The watch manufacturer Omega has 1,000 employees.

At the national exhibition in Bern, the Bieler company exhibits a skeletonized very large pocket watch (52 cm).

Beautiful catalogue with a large wristwatch section (jewelry watches).

1917 The British Royal Air Force gives Omega watches to its airplane crews.

1918 The American Army corps, which is deployed in Europe, is equipped with Omega watches.

1919 The Omega chronometer achieves the highest rating at a precision competition at the Observatory in Neuenburg.

1924 The Geneva Omega factory manufactures more watches than the Bieler main factory, where a restructuring and downsizing of its Caliber program is started.

1925 L. Brandt & Freres is starting a cooperation with Chs. Tissot & Fils in Le Locle in order to combine its research and development in the movement sector.

At a Chronometer competition in Kew-Teddington in England, Omega achieves the first prize with 95.9 out of 100 points. First Prize for Omega jewelry watches at the Exposition des arts Decoratifs in Paris.

1931 Omega can improve all precision records at the observatory competition in Geneva.

1932 Omega is chosen time keeper for the Olympic Games. It is the first time

The industrial area of Omega in 1906: watch manufacture (A), Case manufacture (B) and ateliers (C-G).

217

that one single company is chosen. Over the next 60 years Omega will be chosen 21 more times as the official time keeper of Olympic Games.

1933   Omega once more sets a new record in Kew Teddington. Omega becomes the official chronograph supplier to the Italian Royal Air Force.

1936   In Kew-Teddington, Omega sets the all-time precision record for a mechanical movement (97.8 points) with a pocket chronometer (Caliber 47.7 mm).

1938   Introduction of the famous Caliber 30 mm.

1939   "The Miracle of Precision," the chronometer de bord (Caliber 47.7 mm) is shown at the World Exhibition in New York.
The Royal Air Force starts giving water-resistant stainless steel wristwatches from Omega (Caliber 30 mm) to its troops. By the end of the war, over 110,000 pieces had been ordered.

1942   Introduction of the first Omega wristwatch with automatic winding (Hammer self-winding).

1943   Paul-Emile Brandt starts a social security department for Omega employees. The following year, the pension account is initiated.

1945   At the Ski World Cup in Wengen, Omega uses an independent photoelectric cell for timing for the first time. It replaces the earlier chronograph paired with a finish band.

1946   Use of the first finish camera of modern concept. The later developed Racend Timer revolutionizes the timing of sports events.
Chronometer wristwatch Omega reached 92.7 points in Kew-Teddington.

1947   Death of Gustav Brandt (son of Cesar Brandt).
Omega is the second watchmaker ever to build a tourbillon wristwatch. It is a small series of approximately twelve watches, several of which participated at a chronometer competition in Neuenburg and Geneva and reached top ranks.

1948   For its 100th anniversary the gentlemen's models Centenaire and Seamaster (the later was the sequel model to the military watch Caliber 30 mm).

1950   Omega employs 1,600 people.

1952   The chronometer wristwatch Constellation is launched. It has been Omega's frontrunner ever since.
At the Olympic Games in Helsinki, Omega introduces the time recorder with Quartz technology to sports time keeping. The instrument also prints the exact time.
The Bieler company is awarded the Olympic Cross by the organizational committee of the Olympic Games for its "outstanding efforts for sports."

1954   Paul-Emil Brandt dies on August 25. He is considered the actual architect and master builder of the company in this century.
At the headquarters an ambitious investment and building program is running at this time. At Staemplfli street several buildings are raised to

218

make room for new highrise buildings housing the administrative, finance, and sales departments (1955) and the technical and customer services (1958).

1956 At the Olympic Games in Melbourne, Omega used a half automatic timing instrument with numeric indication (Swim-Eight-O-Matic Timer) for the first time during the swimming competition.

1957 The Speedmaster Chronograph is launched. In 1966 this model is named Speedmaster Professional. It will make history as the Astronaut's wristwatch.

1958 Ernst Brandt (son of Cesar Brandt) dies.

1959 Introduction of the Constellation Caliber 551.
At this point the company employs 2,300 people.
After completion of its 20 million Swiss Frank investment program, the workshops and offices had an area of 33,000 m² (compared to 18,000 m² in 1950 and 12,000 in 1925). The company is constantly growing. In order to find qualified personnel, entire companies are simply taken over.

1960 Omega proves that quality and quantity can go hand in hand: 20,000 Constellation watches, manufactured in one series, obtain a chronometer certificate with honorary mention for their very good results.
The SSIH, a commercial and technical community of interests, is founded and Omega is part of it. The SSIH will consequently gain an important position within the industry.

1961 Omega makes sports timing visible: The Omega Scope allows the time to be shown on the screen.

1962 50.5% of all Swiss chronometers with chronometer certificate are made by Omega.

The main seat of the Omega company with the new mechanic atelier (left) in 1950.

1963   For the first time in its history Omega manages to manufacture 1,000,000 watches in a 12 month period.
        At the timing competition in Geneva and Neuenburg, the wristchronometers by Omega set new records in the categories of individual watch, serial products, and regulation.

1964   Omega, as a three-time winner of the Diamond International Award (1957, 1963, and 1964), is named full member of the International Diamond Academie in New York, the world's highest body in the area of goldsmith crafts.

1965   NASA chooses the Omega Speedmaster as the Astronaut watch for all members of the world space mission.

1967   Introduction of the De Ville series for stylish watches. Elegant Seamaster watches had been carrying this name since 1960.
        The 1 millionth chronometer certificate is issued for an Omega watch.

1969   On July 21 Astronaut Neil Armstrong is the first human being to set foot on the moon during an Apollo XI expedition—and with him, a Speedmaster-Professional Chronograph by Omega.

1970   Omega is awarded a Golden Rose in Baden-Baden for a lady's bracelet watch creation. By 1982, the Omega company will have won this prestigious award six times for design accomplishments.

1972   In order to further develop the timing of sports events, Longines and

The Omega company after extensive extension in the late 1950s with its new buildings on Staemppfli street in Biel.

Omega join forces under the patronage of the Swiss watch industry and found Swiss Timing. The newly founded company is the response to the aggressive Japanese competition.

In order to centralize the company operation, the Geneva manufacturer is moved to Biel.

1974 Market introduction of the electronic marine chronometer for the wrist, the Megaquartz 2400. A general flattening of the business cycle. The mega company Omega with its huge product line is having difficulties. The company becomes smaller, destroys the inventory of unsaleable watches and decreases its product line. Extreme personnel reduction. Centralization to its historic location Biel.

1976 The case company La Centrale is closed. Death of Henri Gerber, who in 1928 became technical director at Omega at age 29 and had a large influence on the development of the company until 1965.

The novelty at the Olympic Games in Montreal was the electronic wristwatch Chrono Quartz, (Caliber 1611), which had dual indication (traditional time, digital chronograph).

1978 At the World Swimming Championship in West Berlin, Omega manages to combine a data processing center with the timing instruments and billboards.

After a general competition by NASA, the manual chronograph Speedmaster Professional once again makes the race and remains the official astronaut watch.

1979 At the Ski World Cup in Val d'Isere, for the first time the rank of the timed participant is shown along with the time.

1980 Omega introduces the ultra thin Dinosaur (Height 1.48 mm) to the market. Another novelty of this year is the Marine chronometer Megaquartz 4.19 MHz.

1981 The diver's watch Seamster 120m is launched.

The Omega Calibers are more and more replaced by ETA Calibers.

1982 Introduction of a Large Matrix Display at the Ski World Championship in Schladming.

Omega sells the Blancpain label, which had been purchased in the 1950s along with the company Rayville, to Jean-Calude Biver.

1983 Merger of the SSIH (association of watch manufacturers) and the ASUAG (association of ebauche makers); the SSIH changes its name to SMH (Societe Suisse de Microelectronique et d'Horlogerie).

1984 Introduction of the Louis Brandt Collection.

1986 Introduction of the product line Art and Symbol.

The Seamaster multifunction in titanium case is introduced to the market.

1988 Diver's watch Seamaster 200m (either quartz or mechanical movement)

1992 Limited edition model for the 50th anniversary of Valiber 27 CHRO.

1993 Mechanical Diver's chronograph 300 m.

221

PRICE GUIDE
by Gordon Converse

The price guide on Omega Watches is based on "Fair market values", what a "...willing buyer and a willing seller" might agree on as a price in the United States. It assumes the watches are in generally good condition. In some cases Omega sells some of these models or very similar models in the retail market, so with these we have assigned values as they might be if offered on a wholesale basis - less than retail - trying to reflect a fair comparison with the vintage watches, which are the majority shown. These prices represent a guide, but like all areas of collectibles and antiques, values are fluid and change all the time for no logical reason and change from country to country, and even region to region. This is especially true with the Omega Watches at this writing. In a recent 2007 auction this appraiser found prices especially high on Omega Watches, like a spike on a graph. Although this appraiser feels that this unusual sale brought up the values of all Omega Watches, we are hesitant to appraise the watches in this book at the same very high levels that some examples brought at this recent auction. Instead we assigned values that were higher than one would ordinarily expect yet lower than the auction results.

Page #	Price	Page #	Price	Page #	Price	Page #	Price	Page #	Price
Cover	$3,000	44	$200-850	99	$1,200		156 $200	136	3579 $600
3	$1,800	45	$200-750	100	$4,500		157 $400		$2,200
6	$20,000-30,000	46	$1,500 (steel)	101	$12,500		158 $800		3580 $600
8	$4,500	48	$500	104	$600		159 $400		3581 $1,500
9	$2,500	49	$650	105	$2,500		160 $800		3582 $1,800
10	$350 (platinum		$500	106	$2,000		161 $1,500		3583 $2,200
	and diamonds)	51	$3,500	107	$2,500		162 $800	137	$300-$1,200
11	$650 (18k gold)	52	$5,000	108	$500		163 $1,200	138	$500-$1,200
	$350 (18k gold)	53	$750	109	$750		164 $2,800	140	198 $800
12	$1,200	54	$3,000	110	$600	132	165 Silver $400		199 White gold
	$1,800 (military	56	$3,500 (RP)	111	$750		166 Plate $500		$650
	with cover)	57	$1,500	113	$1,200		167 $800		200 $1,000
	$3,400	58	$3,500	115	$2,800		168 $800		201 $800
14	$4,500	60	$20,000	117	$2,500 ea.		169 $400		202 $650
15	$1,200 (RP, 1990)	61	$1,800	127	$3,500		170 $500		203 $1,000
16	$3,000 (RP, 1985)	63	$3,500		$3,500		171 $800		204 $800
	$6,000 (RP, 1985)	65	$600		$3,000	133	172 $800		205 $650
18	$750	66	$450		$4,000		173 $800		206 $1,000
	$600	67	$3,500 (RP)		$3,000		174 $800		207 $800
19	$1,800		$2,000 (RP)	129	133 Silver $800		175 $400		208 $650
20	Sold for $28,600	69	$850		134 $1,000		176 $500		209 $1,000
21	$12,500	72	$2,200		135 Gold plate		177 $800		210 $2,500
23	$4,500	73	$4,500		$1,500	134	3569 $1,200		211 $3,500
	$9,500		$3,500		136 $2,200		$2,500		212 $3,500
25	$5,500	74	$25,000 (RP)		137 $900		3570 $1,200	141	178 $2,200
26	$4,800	77	$350		138 $2,200		$3,000		179 $2,200
27	$1,800	78	$350		139 $800		3571 $800		180 $1,800
28	$1,200-2,800	79	$450		140 $2,200		$1,200		181 $2,200
29	$1,200-2,800	80	$350		141 $1,000		$2,800	142	186-189 $800 ea.
30	$1,200-2,200	81	$2,800		142 $2,000		3572 $500	143	182-185 $800 ea.
31	$1,200-2,300	82	$750	130	143 $200		$1,000	144	190-193 $800 ea.
33	$2,500	83	$1,200		144 Silver $350		3573 $3,000	145	194-197 $800 ea.
34	$2,800	84	$950		145 Gold plate	135	3574 Gold plate	146	9 ct. gold $400 ea.
35	$8,500 (RP)	85	$950 ea.		$450		$1,000		solid gold $800 ea.
36	$12,500 (RP)	87	$600		146 $800		$1,200	147	$300-1,000
37	$4,000	88	$1,500		147 $350		$2,200	148	$300-1,000
38	$4,800 (RP)	89	$700		148 $450		3575 $2,200	149	9 ct. gold $400
39	$550		$800		149 $800		3576 $800		15 ct. gold $650
40	$600	90	$700		150 $350		$2,200	151	229 $500
	$600	91	$700		151 $800		3577 $800	152	230 $400
41	$1,200 (steel)	92	$850		152 $350		$1,000		231 $500
42	$1,200	93	$850		153 $700		$2,000		232 $300
	$1,200	95	$800	131	154 Silver $300		3578 Gold plate	158	$300-$400
43	$200-800	97	$2,800		155 $800		$1,200	159	$300-$900

Index

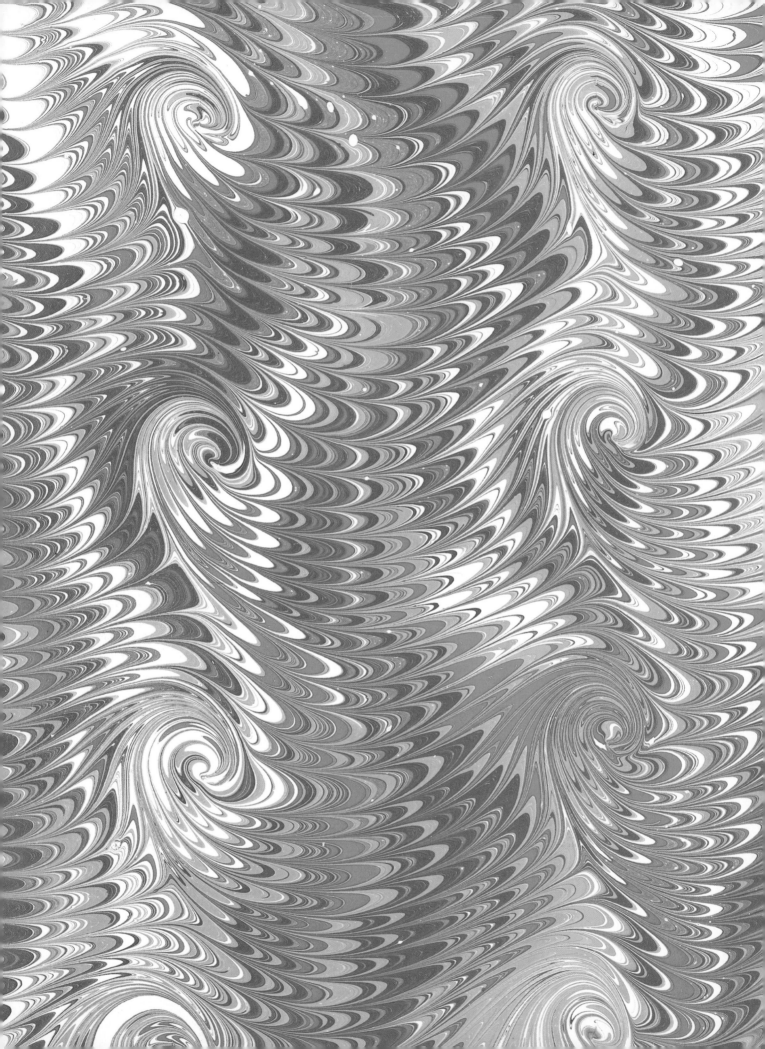